Orchids of Tropical Africa

Orchids
of Tropical Africa

TEXT BY JOYCE STEWART B.Sc.
PHOTOGRAPHS BY BOB CAMPBELL
SOUTH BRUNSWICK AND NEW YORK
A. S. BARNES AND COMPANY

For Heather

Library of Congress Catalogue Number:
70-88583 A. S. Barnes & Co., Inc.,
Cranbury, New Jersey 08512
ISBN 0-498-07555-9 Printed in Italy.

Contents

Preface

The orchids of Africa, particularly tropical Africa, are conspicuously different from those in other parts of the world. From the well-known genera found in other continents many species and glamorous hybrids have been grown in cultivation since the early nineteenth century. Descriptions and photographs of Cattleyas, Oncidiums, Miltonias and Odontoglossums from the American continent and Vandas, Paphiopedilums and Cymbidiums from Asia appear in almost every book on orchids and orchid growing. But until the books by E. A. C. L. E. Schelpe (An introduction to the South African orchids) and F. Piers (Orchids of East Africa) were published recently it was very difficult to find coloured pictures of African orchids. Details about their structure, sometimes only in Latin, could be found by searching in the scientific literature which is available in a Herbarium or University Library. On the subject of the cultivation of all but the most well-known species there was very little to be found at all.

The present work illustrates and describes a selection of forty six orchids from tropical Africa, most of which have been collected during our travels in different parts of East and Central Africa, Madagascar and the Comoro Islands over the last seven years. They have been grown in a variety of situations, both out of doors and in several different kinds of greenhouse, in the Kenya capital of Nairobi. It is hoped that the descriptions of these plants, together with this account of their collection and cultivation, will introduce more of the enchanting and distinctive orchids of Africa and the Malagasy region to orchid growers and thus stimulate an increasing interest in their cultivation.

Acknowledgements

This book could not have been produced without the active help of Mrs. Heather Campbell and Dr. Donald Stewart, who have shared the collection and cultivation of the plants with us. We should also like to thank Mrs. C. Bayer for allowing us to photograph her plant of *Ansellia gigantea*, Mr. C. L. A. Leakey whose plants of *Bulbophyllum falcatum* and *Liparis neglecta* were photographed, and Mrs. P. E. Scott whose Madagascar plants, *Aërangis citrata*, *Cymbidiella rhodochila* and *Graminangis ellisii* also appear in this book. Many members of the Kenya Orchid Society have allowed us to study, and sometimes dissect, their flowers of African species at meetings and shows and we are very grateful to them.

We acknowledge with thanks the help given to us by the Director of the East African Herbarium, Mr. J. B. Gillett, who kindly allowed us to use the Herbarium Library and collections. We also wish to thank Mr. P. F. Hunt, of the Herbarium, Royal Botanic Gardens, Kew, who has confirmed some of the identifications and answered queries.

We are most grateful to Miss Mary Noble of Jacksonville, Florida who read part of the manuscript and made helpful and encouraging suggestions, and to Dr. Stewart who also read the manuscript. We acknowledge with thanks permission to refer to the material published in "Generic Names of Orchids Their Origin and Meaning" by R. E. Schultes and A. S. Pease from the publishers, Academic Press Inc., New York, who hold the copyright.

Introduction

Distribution of orchids in Africa

The family *Orchidaceae* is the largest in the plant kingdom. With 17,000 to 20,000 species, at a conservative estimate, it is much larger than its nearest rival, the *Compositae*, or daisy family. The latter is the larger in numbers of genera, however, with at least nine hundred compared with rather more than seven hundred in the orchid family.

Orchids are distributed throughout all parts of the world where plant life can exist and are most abundant in tropical areas where they form an important part of the vegetative scene, growing as epiphytes as well as terrestrial plants. The epiphytes are merely using other plants, usually trees or bushes, as a means of support, and are not parasitic on them in any way. Their roots extend into the air, among ferns and mosses which also grow on the branches, or over the bark of their support but have no connection with the actual tissues of the host plant.

Approximately one hundred and twenty genera of orchids are now recognised in the flora of the area embracing Africa south of the Sahara, the island of the Malagasy Republic (Madagascar), and the islands of the Seychelles, Comoro and Mascarene groups. Three quarters of these are endemic to this region and the others are mostly larger genera which occur in other parts of the world as well. A total of thirty one genera, twenty of which are found only in this region, are represented in this book.

Although the number of genera in Africa and the islands of the western Indian Ocean compares favourably with other parts of the world the number of species in this area is not so impressive. For such a large area, where a great deal of botanical description and revision is still going on, an accurate estimate of numbers is very difficult to make. It seems likely, however, that there are between two and three thousand species, which is a small number for a whole continent compared with the wealth of species in some of the South American countries: in Colombia, for example, more than three thousand species have so far been recorded. The reasons for this paucity are twofold: firstly, vast tracts of the land mass are unsuitable as orchid habitats, and secondly none of the endemic genera are very large ones. More than seventy genera have less than fifteen species each and about twenty are represented by only one species.

In Africa, as in other parts of the world, the epiphytic orchids are most abundant near the Equator and gradually diminish in numbers towards the temperate zones where they are rare. In the area south of the tropic of Cancer as a whole there are approximately equal numbers of epiphytic and terrestrial genera. The distribution of these within the continent is closely related to the climate, rather than latitude by itself, however, and because most of Kenya, for example, is high and cool, or lower and very dry, the numbers of terrestrial and epiphytic genera are more or less equal, with a much greater diversity of species within the terrestrial group, although the Equator passes more or less through the centre of the country. The Congo, at the same latitude, is also a country of great variety but most of it experiences a much more equable climate with constant high humidity and frequent rain. Consequently it has a greater proportion of epiphytic species in its orchid flora compared with Kenya. The island of Madagascar, although some way south of the Equator, is over one thousand miles long with a great variety of habitats suitable for orchids. Nevertheless the warm, humid climate of the east and northwest parts of the island has favoured the development of epiphytes of many kinds and orchids with this type of growth are more numerous here, and on the neighbouring islands, than terrestrial species.

The parts of the continent which experience a very long, hot, dry season provide very unsuitable conditions for epiphytic orchids. As one moves north or southwards away from the Equator they gradually disappear from all but the most favourable localities. The Republic of South Africa lies partly within the Tropical and partly within the Temperate belt yet of nearly five hundred species of orchids recorded in that country only thirty eight are epiphytes. Furthermore many of these are tropical species which can endure a wide variety of conditions and have extensive ranges throughout the continent.

A very large number of terrestrial species have become adapted to a period of dormancy during the long dry season. Having an underground tuber or rhizome which is protected from the scorching sun they can grow to maturity, flower, and set seed within a few weeks during the course of the rains. Such species are therefore widely distributed throughout the extensive dry parts of the continent although many individual species seem to be somewhat restricted in their range, as are many of the epiphytes. The latter are often confined to quite small patches of forest or single mountain ranges. Other species are found only in certain easily defined conditions of warmth and humidity and reappear in widely separated countries wherever such conditions recur.

Because the conditions they prefer are so important to an understanding of the distribution of African orchids and their treatment in cultivation, we give next a summary of the main climatic types found in the continent and the vegetation they support. This is necessarily very brief but it serves to indicate the tremendous variety of climatic conditions, and therefore habitats, which exists in tropical Africa.

Climate and Vegetation of Tropical Africa

The continent of Africa is shaped somewhat like a pear with its stalk end dividing the Indian and South Atlantic Oceans and its swollen end forming the southern coast of the Mediterranean Sea. It extends about five thousand miles from north to south, reaching the thirty fifth parallel in both hemispheres, and is second only to the continent of Asia with regard to the area it occupies. Because the Equator passes roughly through the centre of the continent a large part of the land surface is located between the tropics and the distribution of climatic and vegetation types is remarkably symmetrical. The plateau of high land which extends southwards from Ethiopia along the eastern side of the continent disturbs this symmetry somewhat as also does the enormously increased area of land in the northern hemisphere compared with the southern. Nevertheless the climate throughout the continent is closely controlled by the sun and its apparent migration between the tropics; hence the divisions between different climatic zones and vegetation types follow the lines of latitude fairly closely.

The climate of Africa is essentially hot and the rainfall unreliable. Most of the countries referred to in this book fall within the Equatorial or Tropical climatic belts (see Map 3) where the average monthly temperature does not fall below 64°F. The movement of the sun brings two maxima of insolation to the equatorial zone and a single maximum to the tropical zones. Temperatures very largely follow the path of the sun, showing slight seasonal differences, and so does the rainfall belt, although this is usually subject to a lag. Broadly speaking, therefore, there are two rainy and two dry seasons in the equatorial zone and a single, longer, wet and dry season in the tropical belt every year.

This seasonal distribution of rainfall, however unreliable in amount, is the chief factor determining the growth of vegetation. The heavy and regularly distributed rainfall of the Equatorial and Tropical Marine climates supports a moist tropical forest often referred to as rain forest. Away from the Equator and the eastern coasts forest growth becomes impossible as the dry season lengthens. Woodland and scrub or scattered trees among tall grasses with a marked seasonal growth cycle become the characteristic vegetation types of these parts. The dependence of plant growth on the availability of rain is clearly shown in Maps 1 and 2.

The Equatorial zone

In Africa a true Equatorial climate affects a zone which extends 5°N and 5°S, and from the edge of the Atlantic Ocean to about 30°E. Temperatures are uniformly high and extremely constant with the thermometer hovering around 80°F throughout the year. Extremes of 100°F and 60°F are rarely encountered and the annual range is usually only about 5°F. The diurnal range is much greater, sometimes reaching 15–25°F and the nights seem pleasantly cool in comparison to the daytime. On the east side of the continent the high land from 30°E eastward brings the temperatures down but the annual range is still small and the regime is essentially an Equatorial one.

Since the temperature varies so little throughout the year rainfall becomes the feature which determines the four seasons in this area. These are alternately wet and dry, or wetter and drier, with the rainfall maxima occurring after the March and September equinoxes. The rainfall is heavy and convectional at both these seasons, frequently falling during the afternoon and evening after a morning of bright sunshine. During the 'long rains' of March–May Nairobi receives an average of 17 inches (Entebbe (Uganda) 25 inches; Lagos (Nigeria) 39 inches in May–July) while in the 'short rains' of October–December only 8 inches fall (Entebbe, 13; Lagos, 16 in September–November). Throughout this zone, however, there is some rain in every month and although the humidity is greatest during the wet seasons it is at all times high, somewhat oppressive for human beings except at Nairobi and other places at higher altitudes, and ideal for plant growth.

The moist tropical forests which are characteristic of the Equatorial zone are evergreen and contain a great variety of trees of varying height. The canopy is very deep, usually starting fifty to sixty feet above the ground and with taller emergent trees extending it to a hundred and twenty or a hundred and fifty feet. Below the canopy a tangle of creepers and lianes is mixed with a dense growth of herbs which makes penetration of this kind of forest extremely difficult.

On the high plateau of eastern Africa at this latitude a distinct sub-type, the High-altitude Equatorial climate, occurs between 5,000 and 10,000 feet above sea level. Temperatures are lower and although the annual range is still small the diurnal range is much greater, as much as 40°F, and the climate resembles that of the tropical areas further from the Equator in this respect. The rainfall regime is the same as that throughout the Equatorial latitudes, although the total is less. Nairobi (5,500 feet) receives an annual average of 39 inches while much of the west African coast receives more than 100 inches.

The lower rainfall and humidity on the plateau precludes the development of rain forest except in small isolated patches. Elsewhere there is a complex mosaic of drier montane forests, deciduous woodland and grasslands of various types which provides an interesting landscape and a great variety of plant species.

The Tropical zones (a) Marine

Tropical Marine climates are encountered from 10°S to 30°S on the eastern side of Africa and on the eastern side of Madagascar. They are distinguished from the Equatorial type of climate by the single maximum of rainfall although some rain, often in the form of mist or drizzle, falls throughout the year. Another feature of this type of climate is the prevalence of violent storms, known locally as cyclones, from December to March which particularly affect Madagascar and the other islands of the western Indian Ocean and sometimes the neighbouring mainland of Africa as well.

At the northern end of this belt temperatures are the same as in the Equatorial climates. At the southern end, in Natal, there are sufficient seasonal differences, up to 13°F in the average temperature, for the recognition of winter and summer.

Along the coast a narrow belt of mangrove vegetation occurs in low brackish areas and along estuaries. Behind this, and extending to the eastern edge of the plateau, is a belt of moist tropical forest. Formerly this was fairly continuous but it has been disturbed by cultivation and commercial forestry activities and now forms a complex mosaic with deciduous woodland and savannah.

(b) Continental

Climates of the Tropical Continental type are the most extensive in Africa next to the Desert type. They are found in a broad zone on either side of the Equatorial type and are distinguished from it by the single maximum of rainfall. A further distinction is that for the rest of the year, often as much as eight months, there is complete drought. Humidity is low during the prolonged dry season and insolation intense, so that very high temperatures, often over 90°F, are recorded. At the same time the cloudless night skies allow considerable radiation and the fall of temperature is rapid. Before dawn the thermometer often records less than 50°F and night frosts are by no means uncommon during the cool season. Temperatures soar higher as the dry season continues, the sun reaches the zenith, and the rains approach. It suddenly becomes much cooler with the breaking of the rains although the increased humidity and the much lower diurnal range of temperature at this time make living conditions uncomfortable.

Nevertheless the rains have a magical effect on the vegetation and landscape. Dry watercourses become flowing rivers and the dusty surface turns into mud or a rich green sward. The scorched vegetation bursts into new green and reddish growth, bulbous plants put up their flowers and the seeds of short-lived flowering plants quickly germinate. In the woodlands and mixed grass woodlands which dominate the scene plants of every kind make extremely rapid growth in preparation for a long period of dormancy during the next dry season.

The length of the rainy season becomes progressively shorter towards the higher latitudes. The amount and reliability of the rainfall also decreases and while 50 inches is common on the Equatorial margin 10 inches is the average amount at the fringes of the desert. Stations in between may have anything between these two amounts and a drought in one year may be followed by floods the next.

The higher altitudes covered by this type of climate, as in the Equatorial regions, experience lower temperatures and a lower annual rainfall. A more or less open woodland with a ground flora dominated by grasses is characteristic of the areas with a pronounced dry season and moderate rainfall. In the drier areas conditions do not support the growth of perennial grasses. An ephemeral ground flora of grasses and herbs combined with xerophytes and succulent plants is mixed with trees, mainly of the genus *Acacia*, in a type of vegetation known as Grass Steppe. This is widespread across Africa at about 15°N, on parts of the plateau of East Africa and southwards in Botswana and the highlands of Madagascar. It grades into a sub-desert Steppe in the still drier areas on the margins of the tropical belt where widely spaced shrubs are the only perennial feature of the vegetation.

Orchid habitats

Some of the many kinds of vegetation found between the varied coastline and the snow-capped mountain peaks of tropical Africa have been mentioned in the preceding pages. Orchid plants are found throughout the entire range, from mangrove swamp and sand dunes to the alpine conditions near the snow. They are distributed through bush, forests and grasslands alike, although many individual species are severely restricted in their range.

The basic types of vegetation are described very briefly below as separate entities: these types are usually clearly defined but are often very limited in extent. Within the space of a few hundred yards one can pass from tall grassland through a belt of woodland into moist tropical forest. As the dominant trees change different conditions are available for epiphytic and terrestrial orchids. We have accordingly described the main features of forest, woodland, savannah, bushland and grassland as they occur in Africa noting particularly the conditions within them which affect the growth and distribution of orchids. With a few exceptions we have limited the examples cited to the genera described in the main part of this book where further details of the ecological preferences of some of the species and their precise distribution are given.

Forests

Many types of forests are differentiated on detailed vegetation maps of African countries but all are composed of tall trees with interlacing branches often several storeys high. The canopy is usually dense enough to inhibit grass growth and the herbaceous plants of the forest floor are all shade lovers. *Calanthe*, *Liparis* and some species of *Habenaria* are among the few orchid genera which prefer this kind of shady and protected habitat.

The rain forests of the Congo and other parts of west Africa are the tallest types with most of their trees a hundred and twenty to a hundred and fifty feet high and with an abundance of woody climbing plants among the tall straight trunks. This type of forest also occurs in Uganda and in a few places along the East African coast as well as in parts of Madagascar and the Comoro Islands. Rain forests are always evergreen because there are always some trees with leaves in the canopy although individually all the trees are deciduous. *Angraecum infundibulare*, *Eurychone rothschildiana* and some species of *Vanilla*, and in the Malagasy region various species of *Aëranthes*, can be found on the tree trunks or among low secondary growth in these shady forests. Most epiphytic orchids are not shade lovers, however, and grow low down only where the sunlight has managed to penetrate a gap in the canopy or where the trees are more widely spaced so that the canopy is less dense. Along the sides of streams, near clearings and at roadsides, many species of *Polystachya* and *Bulbophyllum* are common. They are much more abundant a hundred feet or more above the ground among the branches and twigs of the canopy itself. Some of the more unusual and uncommon orchid species, including *Ancistrochilus rothschildianus* and *Oberonia disticha*, are found here too.

The forests of the East African mountains, particularly those over 6,000 feet above sea level are made up of shorter trees. Many of them also experience distinct seasons as far as rainfall is concerned, with a drier season of two or three months following each similar period of heavy rain. The epiphytic orchids which grow in them are adapted to periods of alternate wetting and drying out. Some of those which occur only in these areas of relatively dry, montane forests have the most extensive aerial roots to accommodate them to this way of life. Several species of *Aërangis* are abundant here in shady places while some of the *Cyrtorchis* and *Tridactyle* species, with their thicker, much more leathery leaves, are found in brighter light, sometimes even in direct sunlight. *Polystachya* and the leafless *Microcoelia* both have their representatives in this type of forest and the unusual *Brachycorythis kalbreyeri* is found in the wetter parts. The 'cloud forests' of the higher and wetter situations, with their abundant covering of epiphytic mosses and lichens on every tree, support large numbers of epiphytic orchids of rather few species of *Tridactyle*, *Bolusiella*, *Angraecum*, *Polystachya* and a few other genera.

Drier forests are found at the margins and lower levels of these mountains, particularly on the western slopes. The prevailing winds, being mostly from the southeast or northeast, have lost a good deal of the moisture they carry by the time they reach the western side of the mountains, but even here there are species of *Chamaeangis*, *Diaphananthe*, *Tridactyle* and *Rangaëris* which appreciate fresh air and a complete drying out in between periodic soakings. In the rainy season these forests are cool, sometimes shrouded in mist like the 'cloud forests,' and the plants are damp all day; but in the dry weather they are exposed to bright sunlight and strong winds so that the humidity is very low except when the temperature falls at night.

Among the terrestrial genera *Habenaria*, *Bonatea* and *Eulophia* are all represented in these drier forests and at the higher altitudes, especially near streams, species of *Satyrium*, *Disa* and *Brachycorythis* also occur.

Woodlands

Woodlands of various kinds are very extensive in Africa and often merge with forests, especially the drier types. They are usually distinguished from them by the fact that the trees are of approximately the same height and spaced so that their crowns touch forming a light but more or less continuous canopy. They are made up of many different species of tree which are usually all deciduous during the dry season so that they are leafless for several months during the hottest part of the year.

Only a few epiphytic orchids can tolerate exposure to strong light for such long periods. *Ansellia gigantea* is probably the best-known species; it occurs on palms and baobabs in some of the hottest and driest parts of eastern Africa. Even in parts of the Luangwa valley, in Zambia, which has a dry season of eight months, it flourishes many miles away from the slightly increased humidity of the river banks. *Acampe pachyglossa* occurs there too, but, as in the other inland parts of eastern Africa where it is common, it is usually confined to trees on or near the banks of perennial rivers. Some species of *Bulbophyllum* and *Polystachya*, which have coriaceous leaves and drought-resistant pseudobulbs, also occur in these woodlands but, like the two species mentioned above, they are usually found on the lowest branches or the trunks of the trees where they can enjoy a certain amount of shade in the hottest part of the year. A few xerophytic species of *Tridactyle* and *Cyrtorchis*, which have thick, succulent leaves, and the leafless genus *Microcoelia* are the only other epiphytic genera encountered in woodlands at low altitudes, while the genera *Jumellea*, *Angraecopsis*, *Rangaëris* and *Ypsilopus* are also represented in woodlands at higher levels as well as in the dry montane forests with which these woodlands merge.

The number of terrestrial species in African woodlands is very hard to estimate but is certainly large. Most of them have underground tubers or rhizomes of various kinds which send up a flowering stem before or during the onset of the rainy season. They have a short life of only a week or two and many of the plants do not bloom every year. Together with the fact that they flower when travel becomes most difficult this means that many species are known from only a single gathering and there are probably others waiting to be discovered. *Eulophia* and *Habenaria* are the largest genera encountered in this type of habitat, where the soil often remains undisturbed for many seasons and the vegetation above it is removed by burning every year. Impeded drainage on small areas within the woodlands prevents the growth of trees and produces a grassy '*dambo*' which has an extremely rich orchid flora.

Savannah

Savannah is a term of Spanish origin used to denote an open mixture of trees and shrubs standing in a tall growth of grass. From the point of view of species composition there are many different types of savannah but they all result in a landscape with a park-like aspect and are subject to frequent grass fires. Terrestrial orchids of several genera occur among the grasses although they are frequently hidden by them and overlooked by collectors. Some of the most attractive species of *Eulophia* are found in these perennial grasslands, often flowering shortly after the vegetation has been burnt when they show up well against the charred earth with its fresh green shoots of short grass.

Very few epiphytic orchids occur in the savannah vegetation except where this merges with woodland or a small strip of forest near a stream. Some of the tougher epiphytes which can endure a partly xerophytic existence can be found here, including *Cyrtorchis arcuata*, *Rangaëris amaniensis* and some of the Microcoelias.

Bushland

Small trees which branch or fork from the base cover great tracts of the continent, particularly in eastern Africa, forming a thicket or bushland vegetation. This is of varying density, sometimes evergreen and sometimes deciduous, and on its drier margins the bushes become more widely spaced forming a steppe and eventually a semi-desert type of vegetation.

Two distinctive orchids, *Polystachya tayloriana* and *Eulophia petersii*, are found only in this type of vegetation. There are a few epiphytic orchids, mostly those which also occur in the woodlands, but they are rare. The terrestrials are fewer than in the woodlands although some species of *Bonatea* and *Habenaria* seem to prefer the shelter given them by low growing *Acacia* or *Commiphora* bushes in the areas with a higher rainfall and are frequent there. A few species of *Eulophia*, and of one or two small, little-known genera, also flourish in more favourable places within this vegetation type.

Grasslands

The pure grasslands, and those of the savannah country, offer a very different habitat for herbaceous plants compared with the woodlands and bushland where grass is also the dominant plant of the ground flora. In the latter there is always a certain amount of shade from the sun and protection from the wind but in the true grasslands most of the associated plants are completely exposed. Nevertheless a wide variety of terrestrial orchids occurs in these grasslands, particularly at the higher altitudes near the Equator and also at higher latitudes further south.

As mentioned earlier most of the orchid flora of the Republic of South Africa is terrestrial and some predominantly South African genera such as *Disa* and *Satyrium* extend their range northwards to the montane grasslands of other parts of Africa. Most of these are extremely difficult, if not impossible, to grow in cultivation, however, and they have not been included in this book.

There are several kinds of grassland where drainage is impeded, either permanently or seasonally, for various reasons, and these all have their characteristic orchid species. Permanently waterlogged conditions give rise to swamps where several species of *Eulophia* thrive. These can be transplanted rather easily to damp situations in gardens or maintained without difficulty in a non-porous container.

The seasonally moist grasslands, which are variously referred to as *vlei*, *mbuga* or *dambo* in different parts of Africa, have a much richer orchid flora and even though such an area may be only a few acres in extent many different species of *Eulophia*, *Disa*, *Satyrium* and other genera can be found there. The nature of the habitat, with its alternating conditions of airless moistness and desiccation, is very difficult to reproduce. The species which seem to prefer it are rather rarely seen in cultivation, at any rate in flower, and seem to be as difficult to transplant as those of the cooler grasslands of South Africa.

Habenaria, *Bonatea* and several species of *Eulophia* are common in the grasslands of the savannah regions but most of these are also rather difficult to bring into cultivation. The *Eulophia* species with pseudobulbs above the ground present few problems and of the tuberous rooted genera *Bonatea* seems to be easier than others. Many efforts with some of the spectacular Habenarias, and Eulophias with chains of underground rhizomes, have been unsuccessful.

Collecting Orchids

Collecting indigenous orchids is a hobby enjoyed by many orchid growers today, not only in Africa. In fact most amateur growers in tropical regions start off with the local species and when their interest and enthusiasm spreads they import exotic species and hybrids from other parts of the world. Flowering orchids are also collected by botanists but usually in very small numbers; their specimens in herbaria increase our knowledge of the variety and distribution of this fascinating family of plants.

The number of plants and species which is removed from wild situations by these two groups are limited and except in the case of plants which are already rarities there is no threat to the continued existence of the species. This is no excuse for denuding the forests, however, and collectors should always be selective, choosing only the plants which are the most suitable size for transplanting. Usually these will be small or medium-sized, well-developed plants. The smallest ones will be left to grow under the best possible conditions, and the largest ones to flower and produce seeds which will ultimately replace the plants that are removed.

There are certain other rules which should be observed by collectors, both local residents and visitors alike. Few African countries have laws prohibiting the collection of orchid plants; however intending collectors should always find out if there are such laws in the countries concerned and obtain permission to collect from the local Department of Forestry or Agriculture if necessary. Most countries have Nature Reserves and National Parks of various kinds, sometimes covering quite large areas. Although originally established to conserve wildlife or forests these are often total nature reserves and no plants may be collected in them without permission from the authorities. But there are plenty of places outside these areas where orchids abound. In many parts of Africa there are settlement schemes, tsetse clearance zones and irrigation projects, where the forest or bush is being removed at a great rate for cultivation. By removing orchids and other rare plants from these areas the collector is active in the cause of conservation, especially if the plants can be propagated from seed in cultivation, and enhances the variety and interest of his own collection at the same time.

Digging up terrestrial orchids presents few problems provided adequate tools, which must sometimes include a pick-axe, are available. The tubers and rhizomes are often deeply situated and any which are damaged will fail to survive. They should be removed with as much soil as possible adhering to them and their roots and placed without disturbance in a tough brown paper bag, or a plastic bag if this is kept open. The worst time to try to transplant ground orchids is when they are in flower, but as they are often very difficult to locate before or after this stage it is sometimes impossible to avoid this.

Epiphytic orchids are much more commonly sought after and are frequently more difficult to obtain. The problem first of all is to reach them and secondly to detach them from the host plant without damaging too many of their roots. Unless one is extremely agile one is limited in many forests to areas where felling is going on and the topmost branches of the trees are brought to ground level. It is only when examining fallen trees that one really appreciates the numbers of epiphytes which occur within the canopy and it is very pleasing to be able to salvage some of those which would otherwise die. Small epiphytes are best collected on the branches on which they are growing and a small hacksaw is easy to carry around for this purpose. Larger plants can usually be detached by hand or with a broad-bladed knife so that at least some of the younger roots remain intact.

Where an exciting epiphyte high above you catches the eye you must either climb the tree yourself or employ a local child to do so. The latter sounds simple but is not really the best technique as it is difficult to explain to the uninitiated, especially when there are language problems, exactly which plants you want and how they are to be obtained and sent down to you. If you choose to climb yourself this can be made much easier by wearing climbing irons. They need to be designed to fit the boot of the climber and a pair made by a small hardware firm for a few shillings have proved most successful on several occasions.

The alternative to climbing is to have some kind of mechanical grab which can be operated at the end of a long pole. Suitable poles can sometimes be cut in the forest but there is never a long enough one around when you most want it and they are apt to be very heavy and unwieldy. Instead we have used a set of light metal poles which interlock and can be added to as the occasion demands. Each one is six feet long and the set is easily carried in an old fishing rod case. In practice we have found more than three of these poles difficult to cope with–they have an annoying habit of coming apart just as one is within reach of the orchid–but an improved design could probably get over this difficulty. The top pole has a forked hook which is not difficult to manipulate and orchids ten to twenty feet above the ground on trunks, and on small leafy branches where they would otherwise have been unobtainable, have been collected with this apparatus fairly easily.

Once the plant has been gathered it is best to trim off dead leaves and old or damaged roots straight away. Only the newly developing roots will be of use to the plant in re-establishing it so there is no point in retaining the old roots, which are likely to be damp, and to be the first parts to cause rot to set in or spread, while the plants are being taken home. Slugs, snails and insects should be searched for and discarded at this stage too before the plants are packed into a basket or carton.

The safest way to move newly acquired plants is to have them loosely but firmly packed in a cool dry container. Newspaper makes good wrapping material as it insulates the plants and prevents one damp or diseased plant from contaminating any others. Plastic bags of various kinds are useful for terrestrial plants and small, delicate epiphytes, provided they are kept very cool and unsealed, and are inspected from time to time. If allowed to become too damp or warm they quickly spell death to their contents.

The wrapped plants, or small branches bearing plants, are most conveniently carried in straw baskets. Locally made ones of various kinds are available cheaply in all parts of Africa. They are light and easily carried in the hand but also fairly tough and cool.

When travelling from place to place by vehicle it is most important that the plants should be kept cool and shaded. They should never be packed in the boot of a car, left unprotected on the deck of a boat or placed on the spare seat on the sunny side of a light aeroplane. It is surprising how much heat is absorbed by a moving vehicle under intense insolation, let alone when the vehicle is stationary. A load of plants, or even a few in a basket, should never be left in a vehicle parked in the sun. If there is no shade for the vehicle the plants must be taken out and placed in the meagre shade the vehicle itself provides, but not forgotten or run over when you move on again.

On a trip lasting only a few hours or over a weekend the important factors are to keep the plants cool and sufficiently firmly packed to prevent them bruising each other if the journey is rough. A collecting trip lasting longer than this has the additional problem of looking after the plants as you go along. In the early days of a two or three weeks safari it is tempting to leave plants until later so that they will have a better chance of survival until you reach home. On the whole it is best to collect the plants as and when you find them for you may well decide to return by a different route or forget the exact locality of something which would have been really worth having.

When operating from a base such as a hotel or camp looking after the plants is not too difficult. Keeping them spread out in a shady place free from visiting slugs and mice is the best procedure, and if they can be sprayed with water several times during the early part of warm days so much the better. Travelling daily from place to place, collecting plants as you go, presents more difficulties and in this case it is necessary to be very selective. Only the most healthy plants at the best stage of growth for transplanting, usually some of the smaller ones, should be taken. At the end of the day, when you are tired and only wanting food and rest, the plants must come first. No matter how many miles have been tramped the plants collected that day must be tidied up and wrapped and those from earlier days inspected and aired before the collector can consider his personal needs. However tempting to do so it is often quite fatal to leave the plants until the next morning.

Most countries in Africa do not restrict the export of indigenous plants. In common with countries in other parts of the world they do have rules regarding the import of plants and usually require that the plants should be inspected by their entomologists on arrival and that the plants should be accompanied by a Phyto-sanitary Certificate from their country of origin. Visiting collectors must find out what the regulations are in their own country before they leave home. They will not find it difficult to obtain plant health certificates, if required to do so, from the Department of Agriculture of the countries where they have collected plants, and can then be assured of arriving home with the plants without difficulty. Failure to comply with the regulations can result not only in delays which may be fatal but even in the destruction of the plants.

Growing Orchids

Many books and articles have been written about orchid growing and we do not propose to duplicate the excellent information they contain. We have decided to include some details which need to be borne in mind when cultivating African orchids, however, for two reasons. When we first started to grow them ourselves we followed the conventional instructions in some of the standard textbooks. These gave excellent results for the exotic species and hybrids which we also grow, but we did not get the best results with most of our indigenous species in this way. By trial and error we have found methods which suit most of the species described in this book, but we are still at the experimental stage with some others. The second reason is that visitors from other continents, who also grow orchids in their homes, have frequently commented that they have not seen certain local techniques used before. It seems that there are certain features of the cultivation of some African orchids, particularly the angraecoids, which need to be emphasised. The definitions given in this section should be borne in mind when reading the cultural notes relating to each species in the main part of the book.

The most important point in orchid cultivation, which applies to the cultivation of many other plants as well, cannot be over-emphasised and it is that the three essentials of culture–light, humidity and temperature–should be in *balance* with each other. To give two examples: *Ansellia* plants from exposed situations in the warm, humid, low altitude regions of equatorial Africa flourish when transplanted to higher elevations, such as Nairobi, where the temperature and humidity are lower than in their natural surroundings, provided they are lightly shaded during the brightest part of the day in sunny weather. They will survive if fully exposed to the sun but in the drier atmosphere the leaves will be very yellow, if not actually burnt, and this is not good cultivation. Besides spoiling the look of the leaves light which is too strong has an inhibiting effect on growth and plants may be stunted and malformed when grown in unbalanced conditions. Similarly an epiphytic orchid brought down to Nairobi (5,500 feet) from about 8,000 feet on Mount Kenya will need to be grown in a shadier position than the tree top where it was collected, to counteract the higher temperatures it experiences at the lower level.

The growth of plants is most rapid where light, humidity and temperature values are high. Many orchids are naturally very slow growing, however, and prefer cooler, less humid and more shady conditions for their healthy development. Nevertheless their rate of growth can often be increased without harm, so that flowers are produced twice a year instead of once, when the plants are removed to suitable, well-balanced conditions in cultivation.

Light

The strong light in which many African orchids grow naturally without burning, and without which they will not flower, is constantly surprising. In all the tropical regions of the world daylight is as strong throughout the year as it is in London or New York during sunny days in June. The only respite comes in the first and last hours of daylight and on cloudy days when the sun is completely or partially obscured. While many people wear shady hats and dark glasses in order to survive comfortably in the tropics there are several orchids which need to be fully exposed to the same light in order to thrive.

In the descriptions which follow *strong light* means fully exposed to the natural tropical light; *light shade* means as much shade as is provided by a tree with a thin canopy such as *Croton megalocarpus*, certain species of *Eucalyptus*, or the silver birch in temperate regions. Such shade is usually transient and the plant is exposed to alternating short periods of direct sun and shade. This degree of shading is not so heavy as *semi-shade*, which describes conditions in which the plant receives about 50% of direct light. This is available in short periods of alternating full light and complete shade. These conditions are fulfilled in lath houses in which orchids are frequently grown in the tropics, or by the use of slatted screens placed over greenhouses at right angles to the diurnal path of the sun. *Deep shade* means a situation in which direct sunlight is never received by the plant and is equivalent to the amount of light received on the forest floor which is protected from direct sunshine by several layers of leaves.

These four values apply to light conditions in tropical regions and growers further north or south will need to qualify them in regard to the conditions they provide because the strength of light available to start with is lower.

Fresh air and humidity

A number of the orchids described in this book grow

in windy and exposed positions and most of them live where the air is fresh with constant breezes even though the latter may be humid. Many growers in tropical regions keep their plants out of doors in the fresh conditions to which they are accustomed. In temperate regions, where a greenhouse is essential at least in winter, maintaining a fresh atmosphere is extremely important. Normal ventilation often provides an inadequate amount of fresh air movement and usually cannot be increased without loss of heat which is provided at considerable expense. In these circumstances, particularly in a small greenhouse, a fan keeps the air moving in winter while the air is cool and damp and, in conjunction with more ventilation, has a cooling effect in summer when temperatures can soar dangerously.

Stagnant, humid air should be avoided at all times. It is particularly dangerous if there is a large temperature gradient between day and night time providing good conditions for the invasion of bacterial and fungal infections. Provided there is plenty of air movement humidity can be maintained at 65–70% during the day when temperatures are in the 70–80°F range; the humidity will reach higher levels at night when the temperature falls, with beneficial results. Without any air movement, and particularly at the lower temperatures experienced during the winters of temperate climates, humidity should be much lower. Generally speaking conditions which feel comfortable to a human being will be suitable for most orchids; thus dry cold is easier to bear, and more successfully endured by orchids, than damp cold.

Temperatures

For some of the plants described in this book we have been able to give exact temperature requirements, but for most of those with a wide distribution we have been deliberately vague. The reason for this is that many of them are adaptable and can accommodate themselves to the situation the grower provides. In the text *cool* applies to the conditions which plants experience when growing at an altitude of over 6,500 feet at the Equator, or lower at higher latitudes, where temperatures may fall to 45°F or less at night and reach 70–75°F during the day; *intermediate* applies to conditions between 3,000 and 6,500 feet at the Equator, where the temperature is in the region of 60°F at night and reaches 80°F or more during the day; *warm* applies to the lower altitudes, where plants live in a temperature of 65°F or more at night with a corresponding amount of heat during the day.

Containers

Before being potted up and mixed with an existing collection newly collected plants must be carefully cleaned, preferably with fungicidal and insecticidal solutions as well as soapy water. Terrestrial orchids and most of the pseudobulbous epiphytes can be grown in clay or plastic pots, in various fibres or composts to suit the individual species, in the conventional way. Many of the African orchids which are classified among the angraecoids do not do well with this treatment, however, and some refuse to grow at all. Several angraecoids thrive in the more open conditions of a rustic cedar or redwood basket filled with osmunda, tree fern fibre or bark. Their long aerial roots need plenty of space and the ability to drain quickly and dry out after watering. For this reason they also do well on tree fern slabs suspended in plenty of fresh air.

In our experience, however, nearly all these species grow best in their natural way, that is, clinging to a section of a tree limb. If they can be collected on a living branch and taken home already established all they need is to be treated with an effective insecticide before being added to the collection. Otherwise almost any durable, non-resinous hardwood will be found suitable as a support. A branch from a living tree is better than a half-dead log from the woodpile and those with a slightly rough bark usually give the best results. Wood with a very thick or rough bark is often but not always, unsuitable as it consists of so much dead material; this will quickly rot, or produce large mushroom-like growths, under orchid growing conditions.

The key to success, as with growing orchids on tree fern slabs, is to tie the plants on to the host branch very tightly in the beginning. All old roots more than a few inches long should be removed and a pad of osmunda or living moss inserted under the young roots. The plant and moss are then tied tightly on to the log with nylon fishing line, copper wire, or some other resistant and durable material which does not offend the eye. Some people use sisal string or raffia twine which has the advantage that it gradually rots as the plant grows on to the log and can be removed easily when it is no longer required. If the timing is not quite right, however, the plants may fall off just when the new roots were about to fix themselves on to the bark and several months' patience is wasted. Holes are easily drilled in the log in order to attach labels and supports, or wires if it is to be hung vertically.

Once fixed on to the log the treatment of the plant for the few months it takes to establish itself is most important. Because its roots, and therefore its absorbing powers, are small and transpiration is continuous it needs to be sprayed with water several times a day to prevent shrivelling of the leaves. Provided these remain firm and glossy roots will soon develop and spraying can be cut down to once or twice a day.

This method of growing orchids has the advantage that a number of plants can be established together producing a very effective display when they flower.

Watering, drainage and resting

Although the majority of orchids are moisture-loving plants their demands for fresh air are at the same time considerable. In their natural surroundings heavy storms soak the plants but the water drains away immediately and as soon as the rain ceases they begin to dry out. The aerial roots, which swell up with water and appear green during the rain, shrink and whiten as air replaces the moisture which is absorbed into the plant from the outer layers of cells.

Heavy watering with quick drainage, and dry periods between successive waterings, will imitate the natural conditions of most species and promote healthy growth. Water should only be supplied at a rate at which the plant can make use of it, however, so that the intervals between watering in a very humid environment, where the roots dry out more slowly, will need to be greater than in a drier one. Every grower must determine how long these intervals should be in the conditions he provides.

The host trees of most epiphytes, like the orchids themselves, grow most quickly and flower during the rainy seasons. After flowering the rate of growth slows down, and stops altogether in the dry season when there is often a complete or partial leaf fall from the trees. This allows more sun and air to reach the epiphytes and slows down their growth in turn. Similarly in cultivation, after flowering and the completion of a new growth, whether it is a leaf or a pseudobulb, many orchid plants need a rest. This can by induced by increasing the light and decreasing the water supply, and is necessary to ripen the new growth and promote flowering during the following season.

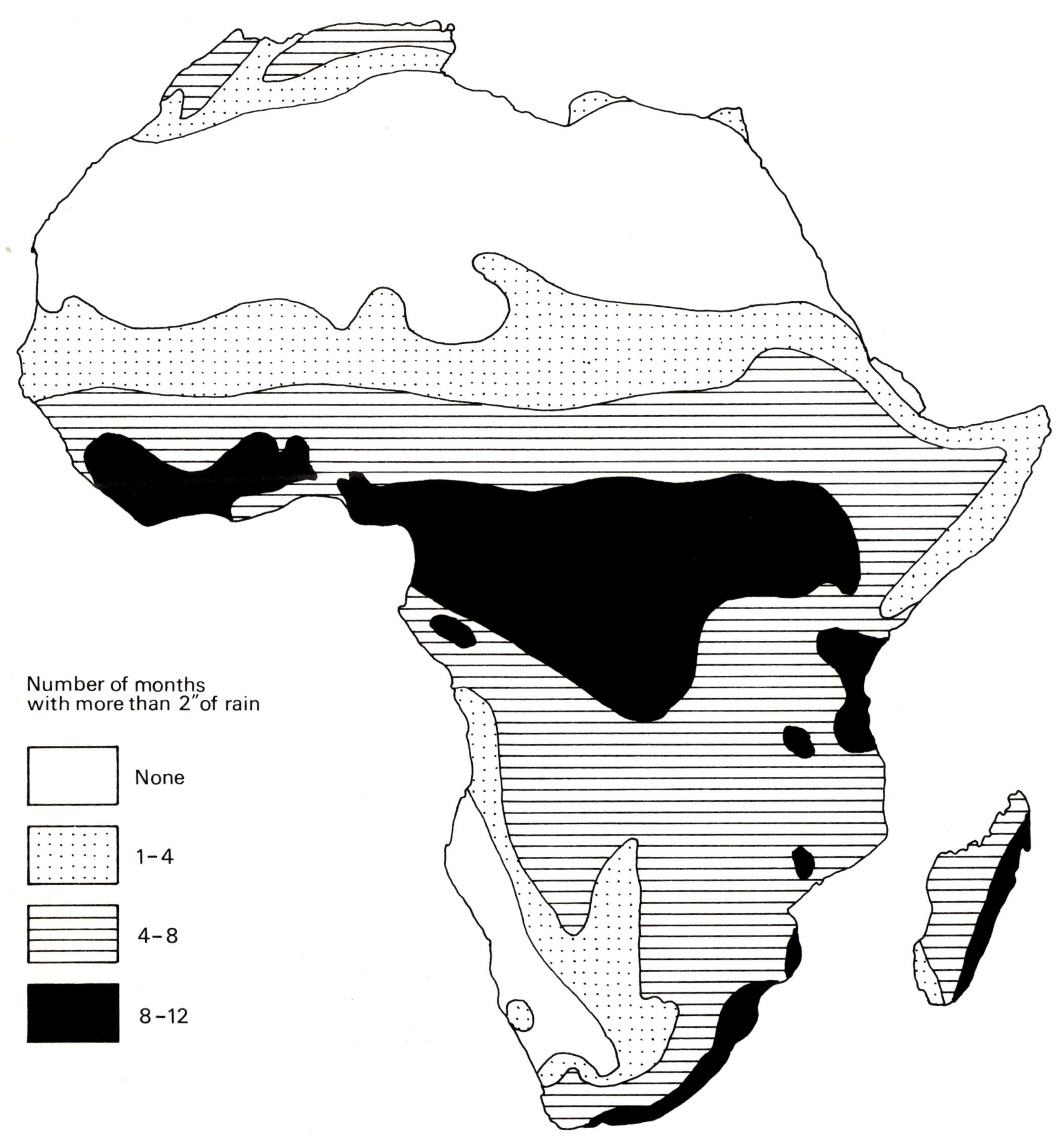

MAP 1. DURATION OF THE RAINY SEASON IN AFRICA

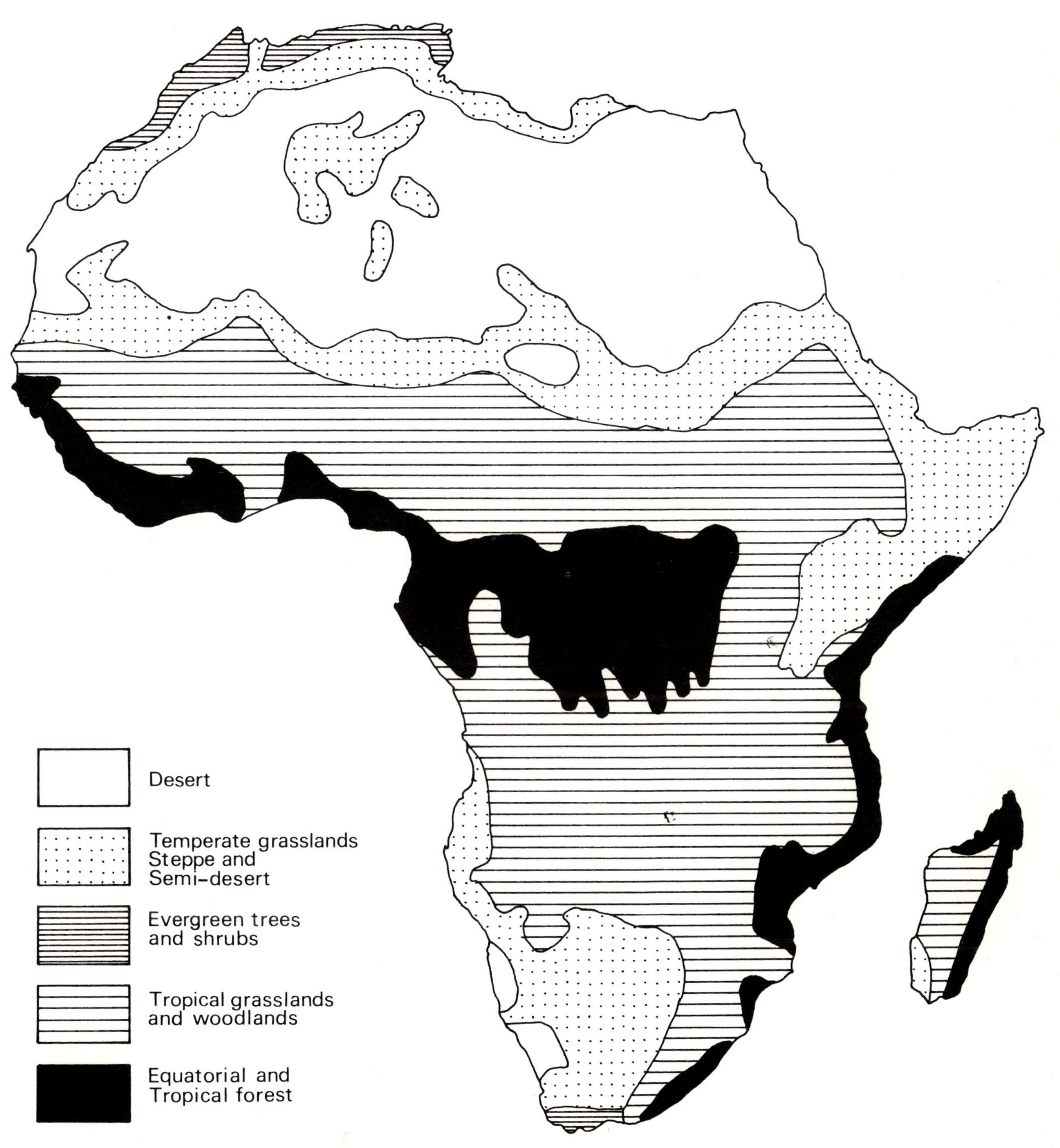

MAP 2. VEGETATION OF AFRICA

MOROCCO
TUNISIA
ALGERIA
LIBYA
EGYPT
SP.SAHARA
MAURITANIA
MALI
NIGER
CHAD
SUDAN
SENEGAL
GAMBIA
PORT.
GUINEA
GUINEA
UPPER
VOLTA
NIGERIA
ETHIOPIA
FR.SOMALILAND
SIERRA
LEONE
LIBERIA
IVORY
COAST
GHANA
TOGO
DAHOMEY
CAMEROUN
CENTRAL AFRICAN
REPUBLIC
SOMALI REPUBLIC
F.PO
RIO MUNI
GABON
CONGO (B)
CONGO (K)
UGANDA
KENYA
RWANDA
BURUNDI
TANZANIA
ZANZIBAR
0°
0°
ANGOLA
ZAMBIA
MALAWI
COMORO Is.
MOZAMBIQUE
MALAGASY
REPUBLIC
S.W.
AFRICA
RHODESIA
BOTSWANA
TRANSVAAL
SWAZILAND
LESOTHO
SOUTH
AFRICA
NATAL
CAPE
100 0 200 400 600 800 1000 ml.
0 500 1000 1500 km.
20° 0° 20° 40° 60°
T. of
Cancer
F
A3
A1
A3
A3
F
A2
A2
0°
0 500 1000 ml.
0 800 1600 km.
T. of Capricorn
20° 0° 20° 40° 60°

ORCHID PHOTOGRAPHY (by Bob Campbell)

For many growers part of the enjoyment of their hobby is achieved in compiling a photographic record of their successes. Flowers produced over one or more years can be displayed in the course of a few hours in this way while perhaps only 10% of the collection will actually be in flower at any one time. For those who are interested in photography, and in particular the close-up photography of flowers, here is a brief description of the equipment and the technique used to take the pictures that appear in this book.

The camera was a Hasselblad 500C. with a 150 mm Sonnar f.4 lens, and extension tubes fitted between the camera body and the lens were used to vary the degree of magnification. Kodak Ektachrome X was chosen for the film material, and finally a Braun Hobby EF.3 electronic flash unit, with a single flash head, invariably provided the main source of light.

The method used for taking the photographs was kept simple. Camera and flashlight were always hand held, eliminating the task of setting up tripods and flash brackets, etc. The aperture selected for the lens remained at f.22 without change, this setting giving a maximum depth of focus without any noticeable effects from diffraction. Except when the subjects were in the open, and natural light could be used in the exposure, the shutter speed was set at 1/125th of a second.

The majority of the pictures were shot indoors. Although it may seem odd, close-up photography in the open presents a number of problems that do not appear in the calm conditions of a closed room. First of all, when extension tubes are used to magnify the image a considerable increase in exposure time is required; for instance, with two 55 mm tubes fitted, the subject in bright sunlight and the lens aperture set at f.22, a shutter speed of 1/8th of a second would be necessary. At this slow speed the slightest breath of wind shaking the subject or even a faint movement of the camera would blur the image on the film. Although the use of electronic flash eliminates this problem to some extent one still has to take care that there is no excessive movement of camera or subject. Then again, the exposure balance between sunlight and flash must be carefully computed to avoid unnatural effects, and on dull days of course the available light is not enough to be of any assistance.

For all the photographs taken indoors the exposure was controlled entirely by altering the distance between the flash head and the subject, and depending on the number of extension tubes in use the measurement varied from a few inches to several feet. The shutter speed of 1/125th ensured that the ordinary room lights used for focusing did not have any effect on the exposure. With the problems of movement and variable light eliminated, concentration could be applied to the need for careful focusing, the composition of the picture, and the avoidance of distracting shadows on the background.

(LEFT) MAP 3. POLITICAL DIVISIONS OF AFRICA
INSET: CLIMATIC ZONES OF AFRICA (A1, EQUATORIAL; A2, TROPICAL MARINE; A3, TROPICAL CONTINENTAL; F, DESERT
REDRAWN FROM A. A. MILLER

ACAMPE PACHYGLOSSA RCHB.F. SSP. RENSCHIANA (RCHB.F.) SENGHAS

Few plants are specially adapted to life in a mangrove swamp and most flowering plants avoid contact with salt water of any kind. There are extremely few orchids which can survive in salty situations but along the eastern coast of Africa there are several species which grow within a few feet of the shore. During storms and at particularly high tides they must be splashed, and sometimes soaked, with salt but without apparent harm.

Several species of *Acampe* have been described from the African coastal habitat but these are now regarded as representing two subspecies of *Acampe pachyglossa*. *A. pachyglossa* ssp. *pachyglossa* occurs on the mainland from Kenya south to Swaziland, both on the coast and in hot areas inland. It is usually epiphytic in light shade or occasionally lithophytic in slightly darker conditions. On some of the small rocky islets along the coast of Kenya it occurs on coral rocks overhanging a sandy beach where its straggling roots are wetted twice daily by the incoming tide.

A. pachyglossa ssp. *renschiana* also grows near the sea in warm areas on the western slopes of Madagascar and in the Comoro Islands. The photograph shown here was taken on the island of Grande Comore. The plant concerned came from a huge colony which was surrounding the trunk of a baobab tree. All the stages of flowering, from tight buds to the 4 cm long, angular capsules, were present at the same time in December indicating that the plant is probably in flower throughout the year. In fact when we saw it the baobab had a full canopy of leaves and there were rather few flowers on the *Acampe*. There are probably many more during the period when the host tree is bare.

Another dozen species of *Acampe* occur in Asia and both the African species and the Asian have very similar flowers. They are all thick and fleshy and suited to their generic name which is derived from a Greek word meaning rigid. The sepals and petals are usually yellow and variously barred and spotted with red. The lip is basically white with coloured markings and a variety of papillae and callosities on its upper surface.

The flowers of the African species are 1·5–2 cm in diameter. The chief difference between the two subspecies seems to be in their foliage which is longer and more flexible in the Madagascan than in the mainland one. Plants from the Malagasy region are also easier to flower in cultivation. Both should receive the same treatment as the warmer species of *Vanda* but plants from the islands do well with less light and slightly lower temperatures than those from the continent.

ACAMPE PACHYGLOSSA SSP. RENSCHIANA PLATE 1

AËRANGIS CITRATA (THOUARS) SCHLECHTER

Small plants which produce large sprays of flowers are very desirable in a collection of pot-grown plants. Many of the orchid species which are endemic to Madagascar fall into this category and one which has been grown in cultivation in Europe since the early part of the nineteenth century is *Aërangis citrata*. The plant is almost stemless and it usually bears four or five oblong leaves. They are never more than 15 cm long and 4 cm wide, unequally bilobed at the tip, and are always a rich dark green with an attractive shiny surface.

It is the long sprays of flowers, often more than one on each plant, which catch and hold the eye. The inflorescence is always an unbranched raceme, 10–30 cm long, and bears 15 to 60 cream or straw coloured flowers. These are 5–10 mm apart so they are nicely spaced on the peduncle and not crowded together. Each flower is 2–3 cm in diameter and has a spur of similar length which is often a dull greenish colour at its slightly swollen tip. The flowers are rarely a true lemon colour, as indicated by the specific name, but can usually be regarded as being tinted with this shade of yellow.

In Madagascar this *Aërangis* is a very common plant where suitable habitats for it still exist. It is scattered throughout the forests on the east coast, from Fort Dauphin northwards, and among the mountains in the centre of the island to an altitude of nearly 5,000 ft. It has a long flowering season, plants having been found in flower in the forests in September and throughout the following months until March. It always occurs in shady places and requires a similar situation in cultivation. As a growing medium both osmunda and tree fern fibre have proved suitable, both in pots or baskets or on rafts of the tree fern. In high humidity but with plenty of fresh air at temperatures in the region of 50°F at night and rising to 70° or 80°F during the day plants of this species grow luxuriantly and flower well.

AËRANGIS CITRATA PLATE 2

AËRANGIS FRIESIORUM SCHLECHTER

Although Portuguese sailors had been exploring the coast of Africa since early in the fifteenth century botanical collecting did not begin on the continent until some two hundred years later. When Linnaeus published his *Species Plantarum* in 1753 very few plants from west Africa were recorded in it although this was about the time that botanical exploration was beginning there. In East Africa, apart from a few specimens snatched by several travellers on short visits to the coast as they passed by on their way to India, collecting did not begin until over a hundred years later. Speke and Grant dried 750 specimens on their intrepid journey from Zanzibar to Khartoum and took them back to Kew in 1863. Their expedition marked the beginning of the exploration of East Africa by European explorers and missionaries who collected plants sporadically as they travelled and worked. With the advent of motor vehicles early in the present century the long lines of porters who had previously been employed to carry goods and specimens could be dispensed with and from this time onwards plant collecting progressed rapidly. Large numbers of specimens were taken to Europe by explorers and botanists and many more were sent back by the resident officials, missionaries and settlers who collected in their spare time.

In 1921 the Swedish brothers Rob and Thure Fries, who were both botanists, led an expedition to the mountain areas of Kenya. Many of the plants they took back to Europe they described and named themselves while some which were named for them were described by other prominent botanists of the day. Dr. Rudolf Schlechter was invited to examine the orchid material they had collected and as a result he named the species illustrated here in their honour and published its description in the notes about their expedition which appeared in 1924.

The specimen on which he based his description was collected on the eastern slopes of the Aberdare mountains. Plants can still be collected at medium altitudes along rivers descending from this mountain range and also between 5,000 and 6,000 feet on the rivers descending from Mount Kenya. The species is also common in the environs of Nairobi, usually below 6,000 feet and in rather dense shade, not far above the ground, in thickets of low trees and scrub. Apart from a wetting in sporadic storms the plants are usually forced to survive a prolonged dry period during the months January to March with a diurnal range of some 40°F. The sudden drop in the temperature during the early hours of the morning means that there is a heavy dew formation which presumably supplies moisture adequate for the plants' needs.

The growth of this species of *Aërangis* closely resembles that of the Asian genus *Phalaenopsis* in size and shape with its fan of four to eight dark green leaves. The racemes of eight to twelve flowers are borne below or in the axils of the lower leaves and usually considerably exceed the leaves in length, sometimes reaching 30–40 cm.

When the flowers first appear they are flat, from 4–6 cm in diameter, and usually a pale green in colour with pinkish brown tips to the perianth. Within a few days they have turned completely white and the sepals and petals become strongly reflexed. The specimen described by Schlechter had a spur 13 cm long and many can be found which exceed this by 3–5 cm. The flowers have a very strong nocturnal scent; beginning with remarkable regularity about an hour before sunset every day the scent from a large group of plants can be quite overpowering. Presumably it is an attraction for night flying moths, or some other insect capable of pollinating the flower, but this has not yet been observed.

AËRANGIS FRIESIORUM PLATE 3

Growing this species does not seem to present any problems. The plants grow most luxuriantly when securely fastened to a piece of hard wood so that their roots are able to spread naturally and dry out quickly after a wetting, but they also grow well in slatted baskets containing fairly open fibre or chunks of bark. Fastened to a tree in a Nairobi garden or in a basket in a shady position out of doors the plants require very little attention and reward one with several sprays of flowers every May and sometimes again in November. In other situations it seems likely that the species will do best treated as a cool growing orchid in heavy shade. A dry period after the completion of the growth of each new leaf is probably necessary to instigate the production of flower buds.

AËRANGIS RHODOSTICTA (KRAENZLIN) SCHLECHTER

Aërangis, *Angraecum*, *Rangaëris*, *Jumellea* and related genera usually have white or greenish flowers, occasional forms being found which have rose to brownish tints in the spur. Because they are relatively easy to grow in cultivation, at least in the tropics, and are very floriferous and common epiphytes they have given rise to the myth, commonly believed in horticultural circles until quite recently, that 'all African orchids are white'. In consequence the first collector of the plant illustrated here was most surprised to find in Uganda an angraecoid epiphyte with a bright red column, and quickly despatched some plants to Messrs. Sander at St. Albans in England who were equally intrigued. They provisionally named it *Angraecum mirabile* on account of this astonishing feature and first exhibited it at one of the Royal Horticultural Society's meetings in 1922. It rapidly became known in Europe, among orchid growers, and was regarded as a striking novelty.

The plants do not seem to have flourished for long in cultivation. However, within a few years E. P. Firth found this orchid on the Thika river near Nairobi not without some difficulty, for it was again located on the small branches of trees overhanging waterfalls. He eventually obtained a plant with five fine inflorescences of white flowers with bright red centres. When some of these plants reached Europe they were found to be the same as the *Angraecum rhodostictum*, named by Professor Kraenzlin from plants collected at Yaunde, in Cameroun, at the end of the last century. This had been transferred to the genus *Aërangis* when Schlechter revised the taxonomy of the angraecoid orchids in 1918.

Since E. P. Firth's expedition the wide distribution of this species has gradually been appreciated. In the gorges of the Thika and Chania rivers near the town of Thika it is now almost extinct, owing partly to the depredations of collectors but mainly to the clearing of suitable forest areas on the river banks for cultivation. Further north it can still be found on bushes growing on the banks of some of the rivers which flow east and southeast off the slopes of Mount Kenya, always very near the river and often on branches overhanging waterfalls. In west Kenya it appears to flourish in a slightly different habitat. There we have found plants growing on the trunks of quite small wild coffee trees on the banks of the Mara river where they would have to withstand long periods of dry weather. Near Kampala in Uganda this species is quite common on bushes in well-established coffee plantations, not always very near water. Further afield it has also been recorded in the Congo and Cameroun and Ethiopia near the headwaters of the blue Nile. In general it seems to be limited to warm moist localities between 4,000 and 6,000 feet altitude and is always found in sheltered and shady positions.

It is a small plant, usually not more than 15 cm across, with three to four pairs of dark green leaves. It is not difficult to find, bearing in mind its habitat preferences, but once collected it is not easy to establish it in cultivation. Some good plants are grown in Kenya collections in slatted baskets on osmunda fibre, and rather better ones on the trunks or branches of gardenia or coffee bushes or on the branches on which the plants had been collected. Quick drainage away from the roots, and a warm moist situation, seem to be the most important factors for successful cultivation.

This species is disappointing, however, in that it frequently grows very well for several years and then suddenly dies. Perhaps this is a protest against a life in captivity but it seems more likely that in any case this is what would happen to it naturally. This idea is supported by the fact that one often finds whole colonies of the plants, all of the same size, indicating growth from a single seeding at some time in the past, perhaps preceding the death of the parent plant. Nevertheless the exceptional beauty of this species is a more than adequate compensation for its short life span.

AËRANGIS RHODOSTICTA PLATE 4

AËRANTHES GRANDIFLORA LINDLEY PLATE 5

The island of Madagascar, now known as the Malagasy Republic, is separated from the African continent by the warm Mozambique channel which is only 250–500 miles across. Nevertheless this separation, coupled with the island's proximity to the warm Indian Ocean and consequent abundant rainfall on its eastern side, has produced many differences in the vegetation. A large number of endemic species and some endemic families are found on Madagascar and the adjacent islands. Several genera of orchids are included in this endemism of which *Aëranthes*, *Cymbidiella*, *Grammangis*, *Neobathiea* and *Oeoniella* are illustrated in this book.

Another feature of the vegetation, which is a direct result of the abundant rainfall, is its luxuriance, in numbers of species and plants, and in particular in the wealth of epiphytes which are found on the forest trees. Every twig and branch and many of the trunks of the moist shady forests support lichens, mosses, ferns and flowering plants. In turn, some of the latter, including epiphytic orchids, also have liverworts, lichens and fungi growing on their leaves. These are very difficult to remove, while they are still alive, without damaging the leaves of the host plant, but after a week or two in drier surroundings they die and can easily be washed off. Some plants of a species of *Aëranthes* which we collected on the island of Grande Comore were supporting such a dense growth of lower plants on their leaves that we were not at first sure that the underlying plant was an orchid. However some adjacent plants were already producing the wiry, pendent flower stems which are characteristic of the genus, some of them over 1 m in length, and we knew that these plants would be worth collecting and cleaning.

Aëranthes grandiflora is one of the more widely known species from eastern Madagascar; in this species the wiry peduncle is usually 10–60 cm long, and entirely covered by brownish, overlapping bracts. The elongated parts of the perianth are tinged with yellow at their tips and the sepals are 5 cm long, the petals slightly shorter. Possibly these long, narrow appendages, swinging in the slightest breeze at the tip of their long, thin peduncle, are so formed to attract the eye of an insect. It is probably a bee, or other insect with a fairly short tongue, which lands on the broad plate of the lip to extract nectar from the club-shaped spur and pollinates the flower while doing so. The flowers have a sweet, but not very strong, scent so it seems likely that it is appearance rather than smell which is used to bring the insect to the flower.

There are about 30 species of *Aëranthes* and all are epiphytic on the trunks or large lower branches of forest trees, usually in rather deep shade. They are stemless or with only a very short, woody stem bearing four to ten rather fleshy or leathery leaves, mostly unequally bilobed at their tip. The inflorescences bear one to several flowers, according to the species, and are usually very slender and pendent, but upright in one or two species. Most of the flowers are in various shades of whitish green or green, although the largest ones, those of *A. henrici*, are sometimes pure white. The characteristic feature of them all is the enormous expansion of the foot which separates the base of the column from the spur. In the species with hyaline flowers the continuous venation from the column, through the foot, spur and lip can be seen clearly.

Most species of *Aëranthes* are easy to cultivate provided they are given good drainage and the one illustrated is no exception. We have grown plants in osmunda fibre mixed with charcoal and in tree fern fibre in both of which they have produced dozens of healthy roots. They need to be watered fairly frequently and grown in the shadiest part of the orchid house in intermediate conditions of temperature and humidity. When the flower spikes begin to appear it is best to put the plants on a high shelf so that the pendent flowers can develop naturally and be viewed to advantage.

ANCISTROCHILUS ROTHSCHILDIANUS O'BRIEN

The origin of the many unusual and diverse names which have been given to orchid genera provides an interesting and absorbing study. A recent book by R. E. Schultes and A. S. Pease is largely devoted to this subject and supplies a great deal of interesting information. Of the 1250-odd generic names quoted in the book the authors have found some 22% which have been formulated to commemorate specific persons. *Ansellia* (Mr. J. Ansell), *Bonatea* (Professor M. Bonat), *Bolusiella* (Dr. H. Bolus), *Jumellea* (Professor H. Jumelle) and *Neobathiea* (M. H. Perrier de la Bathie) are among African genera whose names have arisen in this way.

Some names have been derived from classical figures in history or mythology and others from native languages. *Angraecum* is one of the latter; the name is merely a latinised form of a Malay word *anggrek* or *angurek* which the people of Malaya use for all orchids with thick, spreading aerial roots, particularly for *Vanda* and *Aërides* species which have this habit and which are common in their country. The South American genus *Aa*, whose name is popularly supposed to have been invented in order that it would always be the first in an alphabetical list of orchids, is an amusing one, and not so difficult to spell as some others. Certain authorities now consider this genus should be included in *Altensteinia*, so one of our nomenclatural novelties may be lost.

By far the largest proportion of names are derived from Greek words which are descriptive of a distinctive part of the plant or flower. *Ancistrochilus* is one of these, its origin being in the two Greek words *ankistron*, hook and *cheilos*, lip. The genus was described by R. A. Rolfe and named after the curious lip, somewhat resembling a fish hook in shape, which is a distinctive feature when the flower is viewed from the side.

Three species of *Ancistrochilus* have been collected in tropical Africa but it is not always easy to distinguish between them. They are uncommon but widely distributed in moist forests from Sierra Leone, through Nigeria and Cameroun, eastwards as far as Uganda.

The flowers pictured here were grown in cultivation from plants collected in the Budongo Forest of Uganda. They are somewhat difficult to find being restricted to the mossy branches and trunks of trees in dense rain forest.

The pseudobulbs of *Ancistrochilus* are more or less onion shaped. In each growing season a new one arises near the base of the old, subtended from it by a short rhizome. Young leaves are put up first and these eventually attain a length of 20 cm and a width of 3–4 cm. They are plicate, rather thin, and light green in colour. The new pseudobulb gradually develops within the base of these leaves and when mature one or two inflorescences grow from its base, usually before the leaves have fallen but sometimes afterwards. Two or three pale mauve flowers, 5–8 cm in diameter, are borne on each raceme and each has a deeper mauve flush on the characteristic lip whose side lobes are often greyish.

After flowering it is wise to allow this species, which is best grown in warm humid conditions in a pan or shallow basket, a dry resting period. Then repot it in a fresh supply of a fairly rich but well drained compost of leaf mould and fibre and water abundantly as soon as growth of the new pseudobulbs begins.

ANCISTROCHILUS ROTHSCHILDIANUS PLATE 6

ANGRAECOPSIS GRACILLIMA (ROLFE) SUMMERHAYES

The German botanist Fritz Kraenzlin described the genus *Angraecopsis* in 1900. He gave it this name on account of the similarity of the plants and flowers to the well-known genus *Angraecum* (Greek: *opsis*, appearance). About 15 species are now recognised as members of this genus from various parts of Africa and the islands nearby. They are mostly rather small plants with a short woody stem bearing a few strap-shaped or obliquely curved leaves. The slender inflorescences, with their small and often rather spidery flowers, arise below the leaves on the woody stem and are often pendent.

One of the largest species, *A. gracillima*, is shown on the opposite page and several of the characteristic features of the genus can be discerned. The dorsal sepal stands upright above the column and is only one third of the size of the two lateral sepals which are long and spathulate and project downwards and slightly forwards in a position parallel to the lip and slightly behind it. The triangular petals are strongly reflexed and are joined to the lateral sepals at their lower margins. The lip has three lobes and in this species the middle lobe is about twice as long as the strongly reflexed side lobes. A spur with a rather narrow mouth appears from the back of the lip and is just over 4 cm long. These features can all be seen in the views of different flowers in the photograph and in addition the bright orange colouring at the extreme base of all the perianth parts shows up in some of them.

Professor Kraenzlin actually founded the genus on the slightly larger species, *A. tenerrima*, which is common in the shady forests of the coastal mountains of Tanzania. It has greenish flowers and has also been recorded from west Kenya. It is much rarer there, however, and in fact seems to be replaced in the suitable moist forest habitats of Kenya, Uganda and the eastern parts of the Congo by *A. gracillima*. The latter also occurs at slightly higher altitudes, being found from 4,000 to 7,000 feet above sea level in various localities, while the former is mostly found in the warmer 2,500 to 4,000 feet range. Both species are normally epiphytic on the branches of forest trees and specimens have also been collected from damp rocks near waterfalls, always in shady places.

Plants of both the species mentioned above are easily established in pots of tree fern fibre or on rafts or blocks of the same material. In rather shady conditions, where the daytime temperatures do not exceed 80°F and humidity can be maintained above 50%, the plants grow well and several sprays of flowers are normally produced twice a year. The fibre should be kept moist but well drained all the time and extra water should be provided as soon as the flower spikes begin to appear.

ANGRAECOPSIS GRACILLIMA PLATE 7

ANGRAECUM GERMINYANUM HOOK. F.

ANGRAECUM GERMINYANUM (MADAGASCAR) PLATE 8A

ANGRAECUM GERMINYANUM (GRANDE COMORE) PLATE 8B

More than two hundred species of *Angraecum* are distributed throughout Africa, Madagascar and the Mascarene Islands. Of these the greater number are endemic to a small area or a particular island, more than forty being confined to the Mascarene Islands and the Seychelles; of the hundred and twenty or so in Madagascar only a few are also found in the Comoro Islands, two hundred miles to the west. Again, these tiny islands have several species which are found nowhere else, and others which exist there in a slightly different form.

Angraecum germinyanum is one of those which has been collected both in Madagascar (Plate 8A) and on the island of Grande Comore (Plate 8B). It occurs in several different forms which may be merely different geographical varieties or may turn out to be distinguishable as separate species when more material has been studied. It was first discovered in Madagascar in 1886 by one of Messrs. Sander's collectors, M. Léon Humblot. A plant sent to the Royal Botanic Gardens at Kew flowered there first in 1888 and again subsequently. The specific name had been suggested by Mr. F. Sander in honour of Count Adrien de Germiny, of Gonville, France who was an ardent lover of orchids. When the plant flowered at Kew for the first time it was described by Sir Joseph Hooker who pressed the flower for preservation as the type specimen. This plant obviously throve in the Kew collection for a number of years as other flowers were added to the herbarium sheet at later dates as they became available. The very fine plate in Curtis Botanical Magazine, t. 7061, illustrates the flowers and the glossy green foliage of this plant from Madagascar.

It is very similar to the plants we collected on Grande Comore which we found in forest on high land at the north end of the island at an altitude of about 3,000 feet. During rainy weather this forest is often shrouded in cloud but at other times of the year it is warm and dry. The plants were always pendent and rather straggling on the trunks and branches of well established trees. The Madagascan plants occur in similar situations but extend much higher, up to 6,000 feet in altitude.

In a greenhouse all forms of this species are best treated as cool growing orchids. We have them growing well in osmunda and tree fern fibre in pots under cool and shady but rather moist conditions. The species also does well when firmly attached to a chunk or raft of tree fern fibre suspended in suitably moist, but fresh, air.

Although it is not difficult to grow this species it is not often seen in cultivation. It thoroughly deserves a place in any collection of species on account of its unusual flowers. They are very attractive while still in bud and when fully open there is a delightful contrast in colour and shape among the perianth parts. The almost orbicular, glistening white lip is 2–4 cm in diameter and the very elongated sepals and lateral petals sometimes have slightly greenish or brownish tints. Very long spurs are a characteristic of the genus and this species is no exception; the spur reaches 10–14 cm in length. But attenuated perianth parts at least half as long as the spur are rather exceptional, and this species is in fact one of the most striking of a small group within the genus which are known as the 'Spider Angraecums', as they all have very narrow sepals and petals contrasting with a rounded lip.

ANGRAECUM INFUNDIBULARE LINDLEY

The orchids with the largest flowers in Africa and Madagascar are to be found amongst the Angraecums. All members of this genus have attractive white, yellow or green flowers with a more or less concave lip; *A. infundibulare* is shown here, and with its funnel shaped lip 7 cm long and 5 cm wide is most spectacular.

This species occurs in rain forests from Nigeria and Cameroun eastwards to the Congo and Uganda and on the island of Principe in the Gulf of Guinea. Around the northern shores of Lake Victoria and on some of the islands in the lake appears to be the most easterly extension of its range. In the relatively small patches of thick indigenous forest which still surround some of the swamps at the edge of this huge lake, and in an atmosphere which is almost steamy for much of the year on account of its heat and humidity, one can find enormous plants of this orchid. The mean monthly temperature in these areas is always in the region of 70°F and the diurnal variation is about 15 or 20°F. Humidity is at all times uncomfortably high, usually in excess of 70%.

When one has fought one's way through the shrub and liane layer of the forest to the edge of the swamp, and then through the thick stinking mud and rotting leaf mould in which grow tall Aframomums, *Marantachloa* and *Phragmites*, it seems to be only a just reward to find this orchid within reach on the trunk of a tree, instead of 50 m up as are so many of the Uganda epiphytes. At the end of the dry season, in March, the mud, mosquitoes and other insects are least troublesome but the plant portrayed here was collected in October when mud and swamp water were widespread. Dark coloured ants nearly 3 cm long with a fierce bite proportionate to their size were living among the orchid roots on the tree trunks on this occasion and made collection of the plants more hazardous than usual.

This species can be grown successfully in a pot or basket by using large chunks of tree fern fibre, or lengths of tree fern stem, as a surface for the plant to scramble over. It needs to be kept always in shady and very humid, warm conditions for the leaves to retain their healthy succulence, and a well-grown plant will usually produce one to three flowers each year during May. These have a total length of 15 cm, a most delicate scent, and last for about three weeks.

The plants resemble some of those of the genus *Vanilla* in habit in that they are often rooted or at least based at ground level but are epiphytic in using their rather thin and wiry aerial roots to scramble upwards, sometimes among rocks, but more often using the trunk of a mature tree as support to a height of 10 m or more. When they are as long as this the plants are often branched and it is only the top metre or so which bears oblong, rather succulent leaves. These are bilobed at their apex and arranged alternately in two rows along the rather flattened, brittle stems.

ANGRAECUM INFUNDIBULARE PLATE 9

ANGRAECUM LEONIS (RCHB.F.) ANDRÉ

Tropical cyclones are a feature of the climate of Madagascar and the islands near it during the period December–February. The position of these islands, in relation to the adjacent ocean and continent, is comparable to that of the West Indies in relation to the Atlantic and the continent of America. The two groups of islands are about equidistant from the Equator which is another contributing factor to the origin of the violent storms common to both, which in the West Indies are known as hurricanes. Mauritius and Réunion, being the furthest east of the Malagasy group, are severely affected but even the Comoro Islands, 200 miles west of northern Madagascar, do not escape.

Violent winds and torrential rain lasting for many hours or even two or three days are the chief characteristics of these tropical storms. One would expect the natural vegetation in the areas where they occur to be adapted to survive such hazards, and the unusual structure of *Angraecum leonis* can perhaps be explained in this way.

On the island of Grande Comore, where we searched for this species for several days, it was found in only one small area at about 1,500 feet altitude. The plants were always in fairly exposed positions on the trunks or low branches of a species of *Albizia* and other large trees with a thin canopy. Very tough, wiry roots anchored the plant to the host tree, usually directly on to the bare bark but occasionally through epiphytic mosses and fern bases.

The plants are more or less stemless, bearing four to many bright green leaves with closely overlapping bases. The falcate leaves are up to 25 cm long and have a very curious shape. The upper surface is reduced to a knife edge by the cohesion of its two sides and the resulting fleshy blade appears to be laterally flattened and is about 2 cm wide. Small plants of flowering size usually have four leaves arranged in one plane in the shape of a Saint Andrew's cross. Larger plants have more conspicuously falcate, distichous leaves which curve downwards at their tips and are arranged so that they more or less fill a circle.

Water sprayed on to the leaves by rain, or under cultivation, very quickly runs off their waxy surface and drips from the downward pointing tips. The plant itself often grows away from the host tree at an acute angle so that all the drips fall away from it and there is no waterlogging of its basal parts. The structure and disposition of the leaves thus enables the plant to cope with the excessive rain it receives without succumbing to rot or fungal attack.

For success in cultivation it is necessary to imitate fairly closely the natural conditions of most species. It seems unlikely, however, that the plants of this particular species will require such a severe drenching as they experience in their native habitat. High humidity and plenty of fresh air are essential at all times. Temperatures at night should not fall below 60°F and can be allowed to be high during the day when fairly bright light is also appreciated. Plants grown in a greenhouse in these conditions in Nairobi have grown and flowered well.

The white flowers are so thick in texture as to be almost waxy and they are a very pure glistening white. In shape they are very regular, the overall length being 10–11 cm and the greatest width about 1 cm less than the length. The sepals are slightly longer and narrower than the petals and have a thickened keel

ANGRAECUM LEONIS PLATE 10

along the middle of their outer surface. The concave lip shows a tendency to become three lobed in some flowers, but the lips of others on the same plant are completely regular in outline and 3·5 cm across at their widest part. The spur is curved and twisted, sometimes making a regular S-shape; it quickly narrows below its broad orifice and reaches a length of 10–15 cm. Only in the tip of the spur, in the central column, and in the unusually flattened and winged pedicellate ovary of each flower are there touches of green which accentuate its pristine whiteness. Further attractions are that it is very long lasting and has a very delicate scent, even in the daytime.

ANGRAECUM MONTANUM PIERS

The ridge of the Ngong Hills, southwest of Nairobi, is a windswept area over 8,000 feet above sea level. For part of the year the land is shrouded in low cloud and mists for much of the day and night but during the hot weather temperatures may rise well into the 80's although there is usually a refreshing breeze. In the coolest part of the year there are probably frosts at night and during thunderstorms plants growing there may be bombarded by hail.

The trees and bushes which survive in this cool, moist habitat are very stunted and most of those on the eastern slopes are so pruned by the prevailing winds that their branches grow at an angle following the line of the hill and they have a very windswept profile. Those on the steeper western slopes are sheltered, and have a taller growth, but none of them has a canopy extending above the summit of the ridge.

Despite the hard conditions these bushes and trees have become the host for a large number of epiphytes. Mosses and lichens occur there in great variety and there are also a number of ferns. On lichen covered branches three small orchid species, *Polystachya transvaalensis*, *Bolusiella iridifolia* and *Angraecum montanum* grow in close proximity to each other, and a fourth, *Disperis kilimanjarica*, is often found in a more sheltered position low down on the mossy trunk of the same tree.

Angraecum montanum is a miniature orchid in contrast to the other members of its genus described in this book. The plant itself is only a few centimetres high and when not in flower is difficult to detect among the lichens with which it grows. It has been collected in bloom during the rainy season, in May, although it had been completely overlooked during an earlier visit to the area.

Two or three pairs of leaves are borne at the apex of the short woody stem and immediately below these the flowering spikes appear. The rest of the stem is closely entwined and protected by dark greyish aerial roots among which there are also a few sprays of flowers. Each spike is 3–5 cm long and bears five to seven flowers which are uniformly yellowish green in colour and less than 1 cm in diameter. The perianth parts are somewhat thickened and fleshy and rather similar in shape and size except for the lip, which is deeply concave and ends in a tubular spur less than 1 cm long.

Miniature orchids have a strong appeal to most orchid collectors particularly as they provide an interesting contrast to the more flamboyant and well-known large-flowered plants. This particular miniature seems to be one of those which thrive on neglect as it has grown and flowered remarkably well in a shady part of a Nairobi garden. Although it is heavily sprayed with water two or three times a week in the dry weather it has experienced much higher temperatures and drier conditions than in its natural habitat only a few miles away. In temperate countries it should do well in very cool conditions but, to compensate for the less intense insolation in these regions, in a fairly strong light, even direct sunlight, provided it is protected by adequate moisture from burning.

ANGRAECUM MONTANUM PLATE 11

ANGRAECUM SCOTTIANUM RCHB.F.

Cylindrical stems and terete leaves are well-known features of certain species of *Vanda* which commonly grow in exposed positions with only a fairly high humidity protecting them from the burning rays of the sun. In Madagascar and the Comoro Islands there are several species of *Angraecum* with remarkably fleshy, cylindrical leaves, superficially very similar to those of *Vanda teres*, one of which is *A. scottianum*. It seems likely that in this species too the leaf form is an adaptation to enable the plant to survive in highly exposed situations.

On the island of Grande Comore it occurs in several localities on the western slopes of Mount Karthala at an altitude of 1,300 to 2,000 feet. It was most commonly found on the rather bare trunks and lowest branches of a large species of *Albizia*, but also on other large trees in the area both on bare bark and amongst mosses, ferns, and other epiphytic orchids. Nearly all the plants found were exposed to the afternoon sun, being on the west side of the tree trunks some distance below the shading canopy. They were in an area of very high rainfall, sometimes over two hundred inches per year, which falls mainly in the period November to April although there are storms thoughout the year. Temperatures and humidity are constantly high. Large plants of this species were growing in the same trees and sometimes very close to plants of *Angraecum leonis*. Obviously *A. scottianum* will require very similar conditions in cultivation, though it will be able to tolerate, and may even need, a rather higher light intensity.

Young plants of this species are upright and tufted and there are usually several stems growing from a basal mat of wiry roots. Older plants, and most of those of flowering size, become pendent and straggling. Each plant has eight to thirty narrow, terete, almost cylindrical leaves on the apical part of the stem which is turned upwards. The inflorescence is axillary, near the top of the stem, and the 7–10 cm peduncle is curiously thickened towards its apex achieving, at the point where the flowers are borne, a diameter which is twice that at its base. Sometimes the flowers are solitary and accompanied by bracts and the rudiments of one or more abortive flowers, or there may be two to four on each inflorescence, apparently opening in succession.

The sepals and petals are approximately 3 cm long, rather narrow, and of a slightly grey-greenish tinge at first, becoming white. It is the magnificent lip which is the dominant feature of this flower. It is very pure white, partially surrounds the column at its base, and is more or less transversely oblong in shape, 4 cm across, and with a short apiculus at its tip.

As the plant hangs down from its host tree the flowers are held so that the lip is always uppermost. It narrows quickly into the spur which curves round the ovary and hangs down opposite to the lip. The filiform spur is 10–15 cm long and creamy white with a brownish tinge towards its tip. The column is short and thick with two rather squarish auricles projecting downwards below the rostellum.

This attractive species is not apparently very well known in cultivation although it was introduced into Europe as long ago as the 1870's from the Comoro Islands where it is endemic. It is not difficult to grow in a fairly open mixture of bark or tree fern fibre suspended near the glass in a greenhouse providing the conditions described earlier.

ANGRAECUM SCOTTIANUM PLATE 12

ANGRAECUM SESQUIPEDALE THOUARS

The finest of the Angraecums is undoubtedly *A. sesquipedale*, the 'Comet Orchid', which is shown in the photograph opposite. It is confined to some of the hot lowlands of Madagascar and although first discovered by travellers late in the eighteenth century it was not introduced into cultivation in Europe for over fifty years and first flowered there in 1857.

It is a large plant, often attaining a height of 1 m, with two rows of leaves, closely overlapping at the base where they leave the stem, each 30 cm long and up to 8 cm broad. Thick aerial roots extend for several metres over the bark of the trunk of thin-canopied trees on which the plant prefers to grow. Two to six flowers are borne on an axillary spike with ivory white, fleshy sepals and petals and a greenish spur up to 30 cm long arising from the back of the lip. It is from the size of the flowers that the specific epithet is derived; Du Petit Thouars pointed out when he named the plant that the distance from the tip of the dorsal sepal to the end of the spur may be as much as one and a half feet.

All the angraecoid orchids from Africa and Madagascar have spurs, a common character which seems to indicate some connection between the plants and their environment. Probably it is an adaptation to pollination by long-tongued moths and other night flying insects; the nectary is located in the tip of the spur and many of the orchids in this group (though not all) are scented only at night or dusk.

Charles Darwin spent a great deal of time studying pollination mechanisms in orchids and his book 'The various contrivances by which orchids are fertilised by insects' (1862; second edition, 1877) is one of the classics of botanical literature. When he saw the flower of *Angraecum sesquipedale* with its very long spur and nectar filling only the tip, he tried (as he had already done with many other orchids) to imitate the actions various insects might make to reach the nectar and in so doing to remove the pollinia for fertilisation of another flower. After much experiment with bristles, needles and glass rods poked into the nectary he came to the conclusion that pollination of this species probably could be effected only by a large moth with a very long tongue which was proportionately broad at its base. In forcing its tongue to the extreme apex of the spur for nectar this moth would need to insert its proboscis through a cleft in the rostellum as this would be the shortest route to the end of the spur. On withdrawal the tongue would be clasped by the margins of this cleft, the viscid caudicles on the underside of the rostellum would become attached to it, and so the pollinia could be withdrawn from the flower. Every time Darwin tried this with a suitably sized glass rod he was successful and he was also sometimes successful in transferring these pollinia to the stigma of another flower. He was convinced that the development of the very long nectary must have taken place, by natural selection, at the same time as the development of a very long-tongued moth, and that for their mutual benefit each still flourished in Madagascar.

Entomologists were not alone in laughing at this idea, but in fact by the time the second edition of his book had been published in 1877 a sphinx moth with a proboscis ten to eleven inches long had been found in south Brazil, and it was not long before one with a twelve inch tongue was captured in Madagascar as Darwin had predicted. To commemorate this unusual situation, in which an animal's existence was foretold, it was suitably named *Xanthopan morgani praedicta* when it was named in 1903.

Angraecum sesquipedale is an orchid which responds well to cultural conditions, particularly if it is potted in a fairly large container and allowed to remain undisturbed. Pieces of crock and charcoal mixed with chunks of bark and tree fern fibre seem to be the best potting media and the plant should be grown in intermediate conditions of temperature and humidity with plenty of fresh air.

ANGRAECUM SESQUIPEDALE

PLATE 13

ANSELLIA GIGANTEA RCHB.F. VAR. NILOTICA (BAKER) SUMMERHAYES

The 'Leopard orchid' of East Africa is almost as well known to horticulturally minded residents of the area as the animal whose name it bears is to everyone. Like the coat of the animal, however, the coloration of the flowers varies tremendously and in no two individuals, even on the same plant, are the markings exactly alike. Only a few of the 5 cm diameter flowers are shown in the accompanying photograph but these confirm this observation. Plants bearing more than one hundred flowers at one time have been examined and none of the flowers has borne precisely the same markings.

Black leopards are found in the mountainous areas of East Africa and in some areas there are leopard orchids which are almost black too. In these the spots are of a very dark brown or liverish shade and so large that they more or less coalesce to form a very dark flower with indistinct yellowish-green markings. In Rhodesia and South Africa smaller-flowered plants occur whose petals show no markings at all, and these are found in several shades of yellow.

Despite the wide variation in colour and markings of the flowers of Ansellias it seems likely that there are only two distinct species. These are widespread throughout continental Africa south of the Sahara.

Ansellia africana is the darker form and is more typically west African although it extends its range eastwards as far as central Kenya. *A. gigantea* var. *nilotica* occurs in northern Nigeria but is typically the eastern form, being found in Uganda and from Kenya southwards to Zululand. *A. gigantea* proper is restricted to the southern Transvaal and to parts of Natal and Swaziland. Usually *A. africana* is much taller in growth, the pseudobulbs reaching a height of 1 m or more, while both forms of *A. gigantea* can produce huge clumps of pseudobulbs which vary in size from a few to 60 cm long. Plants grown in cultivation do not always show these differences.

Leopard orchids are easy to grow in large baskets, tubs, or even among stones and compost in a tropical garden. They require quite a large amount of space as specimen plants can reach a diameter of a metre or more. When they do so, bearing ten to thirty panicles of a hundred flowers or more each, one panicle from the apex of each pseudobulb, they are a most rewarding sight. A clump of this size might weigh several hundredweight and would be very difficult to move.

In cultivation out of doors in Nairobi this species seems to do best if it is provided with light shade during the hottest part of the day. In its natural habitat it also does best in lightly shaded conditions, although plants have been found in the dry Rift valley and the hotter, drier regions of northern Kenya and Uganda in positions where they are completely exposed to the burning rays of the sun throughout the year. In a greenhouse in a temperate climate they would need to be in the lightest part of the intermediate or hot house, particularly during the winter when days are short, and would probably benefit from a period out of doors in fresh air and strong sunlight during the summer.

The spindle-shaped pseudobulbs are reminiscent of bamboo in colour and markings and each bears three to six pairs of leathery leaves at its apex. These are shed within a year or two of flowering but the pseudobulbs persist for many years, gradually becoming more furrowed as the carbohydrates they contain are used up. One wonders if they also contain any alkaloids or other substances with truly medicinal properties. The Zulu people use a decoction of the stem of the *Ansellia* which grows in their country as an aphrodisiac, an antidote to bad dreams, and for various other purposes; in the Mpika district of Zambia an infusion of the leaf and stem is used to treat madness. In East Africa we have not heard of any medicinal uses but a gum is obtained from the pseudobulb which some tribesmen use to stick the flights on to arrows.

ANSELLIA GIGANTEA VAR. NILOTICA PLATE 14

BOLUSIELLA IRIDIFOLIA (ROLFE) SCHLECHTER

Bolusiella is an easily recognisable genus of about six species of miniature orchids found in various parts of Africa. All the plants have rounded, succulent leaves distichously arranged on very short stems so that they are shaped like a fan. The dressmaker's pin in the photograph gives an idea of the size of one of them, although plants of this species are often twice as large as this one. The plants are anchored to the substratum, in this case a thick branch of a small bush, by thin wiry roots, but they have proved easy to transplant into cultivation, either on to small pieces of hard wood or into pots of tree fern fibre.

The inflorescences are often rather flattened although not always in the same plane as the plant. Alternating rows of brownish bracts partially hide the flowers on each side of the thickened peduncle, and one needs a lens to see that the flowers are really those of an orchid complete with a short spur hidden inside the bract. These flowers are decidedly small and they probably compete with those of *Oberonia disticha* (page 95) for the distinction of 'smallest African orchid'. In fact the flowers of *Oberonia* are slightly smaller in the specimens we have seen but the plants of this species are much larger than those of *Bolusiella*.

B. iridifolia grows in very harsh conditions of wind, cold and damp at high altitudes in Kenya and may even withstand frost on the coldest nights of the year. Cool, well ventilated conditions are therefore preferable in cultivation although the species can grow happily in warmer places. In the Budongo forest of Uganda slender plants of this species grow near *B. imbricata*, a more succulent species, which is more widely distributed throughout the warmer and more humid forests of Africa from Ghana to Mozambique.

BOLUSIELLA IRIDIFOLIA PLATE 15

BONATEA STEUDNERI (RCHB.F.) DUR. & SCHINZ

When collecting terrestrial orchid plants we have often been asked by friends accompanying us 'How do you know it is an orchid?' The answer usually contains an explanation about the floral structure of orchids which is easy or difficult to make according to the plant which is being collected at the time. Somehow one is never asked the same question about epiphytic orchids; perhaps their mode of growth already makes them sufficiently different in the eyes of the uninitiated that they are accepted as members of one of the most interesting and unusual of the flowering plant families without further explanation.

Some terrestrial orchid plants, such as most of the species of *Eulophia*, have very regular flowers whose structure is easy to demonstrate. The three sepals, which form the outer ring of the perianth, are often different in colour and shape from the three petals, and of the latter the lip is conspicuously different from the other two, but still quite obviously, basically, a petal. In the centre of the flower is the column, a more or less massive structure, which houses the reproductive parts, the pollinia and the stigmatic surface. It is directly in front of the ovary which is situated behind or below the perianth and is often indistinguishable from the flower stalk, or pedicel.

In the flowers of *Bonatea steudneri*, and the other members of this African genus, the different floral parts have become so fused together that it is difficult to distinguish them at first sight. Nevertheless by starting from the outside all the parts mentioned above can be identified, and this is especially easy if there are some buds as well as open flowers at the tip of the leafy stalk.

The dorsal sepal is green and convex, almost boat-shaped, and 2–3 cm long. The lateral sepals are greenish white and often curled back on themselves so that their shape is difficult to determine. On the top surface they have become joined to the lip which lies above them. The lip is white and about 3 cm long in its basal part before it divides into three green lobes. The central one is about 3 cm long and 3 mm wide while the outer two are 5–8 cm long and only about 1–2 mm wide. From the back of the lip a slightly sinuous spur, which gradually thickens towards its tip, descends to a considerable length—up to 23 cm in the specimen photographed. The other two petals are divided into long narrow lobes almost from the base. The anterior lobes project forwards on either side of the lip and are also filiform and 5–7 cm long, while the posterior lobes stand upright and curve round the edge of the dorsal sepal.

In the centre of the flower the different parts of the column are much more clearly defined than in many orchid species. The erect anther with its two yellow pollinia can be seen in the photograph inside the hooded dorsal sepal. The stigmatic surface is very sticky and is located on the white, club-shaped tips of the stigmatic arms which are about 3 cm long and which, like the anterior petal lobes and the lateral sepals, are joined to the lip for part of their length. The curiously shaped rostellum in the centre of the column has a helmet-shaped central lobe and two slender side lobes containing the very fine caudicles of the pollinia.

The plant illustrated was dug up in grassland at the side of the road near Thika in Kenya. Planted in a pot in a sandy compost, and kept in a semi-shaded part of the garden, the tuberous roots have survived for several years and put up a leafy shoot at the start of each rainy season in November. This continues to grow, sometimes to a height of a metre or more but usually less, and five to thirty flowers are produced at its

BONATEA STEUDNERI PLATE 16

apex in May or June. Water is given throughout this period and until the leaves begin to go yellow after flowering; it is then withheld until growth begins again the following season. It has been found that weak liquid manure, supplied when the leafy shoot is a few inches high, stimulates growth and possibly increases flower production. Although it has similar tuberous roots this species has proved much easier to grow in cultivation than members of the genus *Habenaria*, which it somewhat resembles, and with which it was formerly classified.

B. steudneri is one of the largest members of the genus and is distributed throughout eastern Africa. It was first collected in Eritrea and is now known to occur in all the countries between there and Rhodesia and to extend its range northwards into the Yemen. It is usually found at the edges of thickets, among grass and herbs half hidden under bushes, but has also been collected on rather bare rocky slopes and in well drained positions among rocks in rain forest; its altitudinal range is also wide as it has been recorded from 3,000 up to 7,500 feet above sea level.

BRACHYCORYTHIS KALBREYERI RCHB.F.

The genus *Brachycorythis*, which consists of about thirty two species, was first described by John Lindley in 1838. He gave it this name on account of the hooded appearance of the perianth, which sometimes resembles a helmet (from the Greek, *brachys*, short and *korys*, helmet). The unusual and rather rare species illustrated here was named by the younger Reichenbach after the collector E. Kalbreyer, who first discovered it on Mount Cameroon.

It is unusual in that it appears to have developed a semi-epiphytic habit. Other members of the genus are straightforward terrestrials but *B. kalbreyeri* is found among moss and ferns on old logs and trees in thick forest near streams, sometimes on decaying timber on the ground, as in the Congo, or actually on old mossy trees as in Kenya in the forests near Kericho. In many parts of its range it undergoes a marked resting season when all the foliage dies back leaving a rhizome rather like that of some ferns just under the surface of the litter. It remains in this state during the long dry season and as soon as there is sufficient moisture the rhizome puts up a new aerial shoot which will eventually bear flowers, and hair-like roots develop on the new growth of the rhizome below the surface.

This species is not difficult to grow in cultivation provided it has shade and a moist atmosphere during the growing period and is later allowed to die back and dry off completely as it would do in natural conditions. It has flowered annually in Nairobi for some years in a basket with mosses and ferns on a surprisingly shallow layer of bark and leaf mould. Each leafy stem grows to a height of about 30 cm and bears up to fifteen flowers, each of which is nearly 5 cm in diameter. The flowers last for several weeks and have a spicy and not unpleasant scent. In general growth the plant resembles some of the Indian species of the genus but its floral structure is much more like a more widely distributed African species, *B. ovata*, which can be found from the Cape Province of South Africa to the Sudan and across the continent to northern Nigeria.

Brachycorythis kalbreyeri now appears to be fairly widespread in the Equatorial region of Africa. Since its discovery in 1877–78 it has been collected from similar habitats in Sierra Leone, Liberia, Cameroun and Congo. It has also been known for some time in highland Kenya in several humid forests near streams above 6,500 feet, but it was not known in Uganda until 1958 when it was collected in a similar habitat in the Impenetrable Forest of Kigezi District.

BRACHYCORYTHIS KALBREYERI PLATE 17

BULBOPHYLLUM FALCATUM (LINDLEY) RCHB.F.

Bizarre and unusual structures are common among members of the orchid family. The plants themselves can all be classified as perennial herbs of relatively slow growth but after that diversification begins. The habit of the plants covers a wide range including climbing, leafless Vanillas, the small leafy herbs such as *Cheirostylis* and *Zeuxine*, the monopodial type of growth of the *Vanda* tribe, including *Angraecum*, and the pseudobulbous epiphytes of which *Bulbophyllum* is an example.

Bulbophyllum is a very large genus, equalling or perhaps surpassing *Dendrobium* for the position of largest genus in the family. Nine hundred or more valid species have been described, all occurring in the tropical and subtropical parts of the world. The plants always have pseudobulbs placed close together or at intervals along a thin woody rhizome and each is surmounted by one or two leathery leaves. The inflorescences always arise from the base of the young or newly matured pseudobulbs. These are characters which hold good throughout the genus and make it an easy one to recognise. When, however, one starts to consider the flowers and their arrangement great diversity is apparent. In some of the East Indian species the flowers are solitary and very large. These have been cultivated in collections of orchid species for many years. Only a few African species have gained such popularity, and that on account of their unusual form rather than their decorative value.

Bulbophyllum falcatum belongs to the section *Megaclinium*, which is described in more detail on the next page, and has a wing-like rhachis about 10 cm long and 8–10 mm wide. The flowers are borne alternately along the centre of each side. The petals are very small, almost thread-like, and difficult to distinguish when one first dissects a flower. The lateral sepals are dark red and almost triangular in shape. It is the single dorsal sepal which is the most conspicuous part of the flower. It is developed into a tongue-shaped structure 8–9 mm long with two yellow, thickened areas at its apex. Ten to fifteen flowers are borne on either side of the greyish green rhachis and make up an unusual and not unattractive inflorescence.

In cultivation *B. falcatum* is easy to grow. If several plants are established together in a basket filled with bark and fibre, or on a chunk of tree fern, they will soon grow into a large mat which produces several inflorescences at each flowering. Warm conditions of semi-shade and high humidity are necessary to produce such good growth as the natural distribution of this species extends from Sierra Leone through the equatorial tropics of west Africa to the Congo and western Uganda.

BULBOPHYLLUM FALCATUM PLATE 18

BULBOPHYLLUM OXYPTERUM (LINDLEY) RCHB.F.

The *Megaclinium* section of the genus *Bulbophyllum* is an exclusively African one of unusual and interesting structure. It was originally established by John Lindley as a separate genus, the name being descriptive of the wide, flat, often slightly spiralled inflorescence in which the flowers are set or embedded (Greek, *megas*, wide or large; *kline*, bed), but botanists have since decided that it is more suitably treated as a section of the larger genus *Bulbophyllum*. Less than half of the African species of this genus belong to this section but those that do are easily recognised as soon as they come into flower.

For the first half of its length the flowering stalk, or peduncle, has the usual round cross-section with a few bracts sheathing it. Above this, on either side of the midrib which bears the flowers, the rhachis is expanded into a leaf-like wing. This is usually thick and fleshy, often gradually narrowed towards the base and tip and twisted spirally so that it is far more spectacular than the minute flowers it bears. The species with the longest rhachis is probably *B. purpureorachis* from the Congo, in which it is 30–75 cm long, 5 cm wide, and twisted through a complete circle. The flowers are rather insignificant along the centre of each side of this growth and rarely open fully.

B. oxypterum from the Kenya coast and several other parts of East Africa is not so large as this but has similar form and coloration. This is seen to advantage with the aid of a powerful lens or close-up photography as the flowers are only 5 mm or so in diameter. They number twenty to fifty on each side of the thickened rhachis which may be up to 25 cm long on a stalk of nearly equal length. At first sight one only appreciates the mottled surface of the rhachis but close investigation near its apex usually yields one or two open flowers whose distinctive and intricate colouring and markings are worth observing. The mottled lip is particularly interesting in that it is hinged to the base of the column and is extremely motile. The slightest puff of breath or wind sets it in motion, and this movement, in addition to the faint but distinct carrion scent of the flowers, probably attracts the small flies that are capable of pollinating this flower.

The sharply four-angled, yellowish green pseudobulbs of this species are about 8 cm high but are rather widely spaced—up to 10 cm apart—on a woody branching rhizome. This makes it a slightly difficult plant to accommodate in cultivation. If grown flat it is best to provide a large square or oblong basket filled with bark and fibre; or the plant can be grown vertically attached to a raft of fern fibre or a suitable piece of hardwood. The chosen container needs to be kept in lightly shaded, warm and humid conditions. In the wild state plants are not usually found above 5,000 feet, although they are widely distributed in suitable places below this altitude from Kenya southwards to Malawi and Mozambique.

BULBOPHYLLUM OXYPTERUM PLATE 19

CALANTHE VOLKENSII ROLFE

The genus *Calanthe* was first described by Robert Brown and now contains about one hundred and twenty species which are distributed mainly through tropical Asia with a few in Australia, Africa and tropical America. The genus is readily divisible into two main types from the standpoint of vegetative structure and cultivation.

In one kind a new and large pseudobulb is produced each year from which the leaves are deciduous. This section includes the best-known species which occur in India and southeast Asia and have been grown in cultivation for many years. The African species all fall within the second group with their strongly ribbed perennial leaves and non-existent or very small corm-shaped pseudobulbs.

The deciduous plants all require fairly strong light for the complete development of each new growth every year. The evergreen species, at least the African ones, will only do well in conditions of fairly dense shade. In their natural habitat they are terrestrial plants on the floor of tropical forests with several layers of leaves, and branches of bushes and trees, protecting them from the sun. In the dense forests covering the steep slopes of several mountain ranges in Kenya, mostly over 7,000 feet altitude, large colonies of *Calanthe volkensii* flourish in the greenish sunlight which filters down to their level. In the primary forests of Madagascar and the Comoro Islands a slightly smaller species, *C. silvatica*, grows in very similar conditions. The plants grow in a very well-drained compost, which is chiefly leaf mould in various stages of decay, mixed with rock fragments and sandy soil. In cultivation they do well in a similar mixture provided they are kept in a shady, cool and moist situation. At the same time the air should be fresh. Otherwise, in the high humidity, the rather thin leaf blades become very susceptible to fungal attack and although this can be kept in check black spots spoil the look of a plant whose leaves are perennial.

Calanthe volkensii flowers in the period March to June and the flowers last for a month or more. The rather open raceme bears twenty to thirty flowers and may be as long as 50 cm. Although slightly shorter than the rather broad, plicate, dark green leaves, the raceme stands well above them as they bend over towards the ground.

Individual flowers are about 5 cm in diameter and a very pretty shade of pale purple with the lip a slightly darker shade of the same colour. A delightful feature of this four lobed lip, however, is that as it ages it changes colour, gradually acquiring an admixture of yellow until it reaches an almost apricot hue before the flowers fall. The column is always white and is fused to the basal part of the lip making a tunnel-like entrance to the nectar-containing spur. Immediately in front of it various callosities have developed which are the same colour as the lip in the Kenya species, but in *C. silvatica* are more extensive and a bright orange yellow in colour.

CALANTHE VOLKENSII PLATE 20

CHAMAEANGIS ORIENTALIS SUMMERHAYES

Many African epiphytic orchids make up for the small size of their flowers by the number they produce on each spike, and sometimes also in the number of spikes produced at each flowering. Even if their structure is difficult to elucidate without recourse to a lens or microscope they are often very fragrant and sometimes have distinctive colours.

Chamaeangis orientalis has all these attributes and the additional unusual feature that the flowers are nearly always upside down on the curved or upright inflorescences, with their 15–20 mm spurs pointing vertically upwards. Two to four flowers are borne in whorls about 1 cm apart along the rather thickened brownish peduncle which may be as long as 30 cm and is usually slightly longer than the leaves. In colour the flowers vary from brownish yellow to a dull orange or apricot shade and their scent is strong throughout the day.

When not in flower the plants are somewhat difficult to distinguish from those of the related species *C. vesicata* which, however, always has yellowish flowers with a distinctly bulbous tip to the spur. The woody stem of *C. orientalis* has long aerial roots below the thick and fleshy leaves. These hang downwards in two rows and are often curved slightly so that their tips are almost touching. They are rather a bluish green in colour often with brownish red tints, and nearly always look rather shrivelled and droopy except during the height of the rainy season.

This species has now been collected from most of the higher mountains in East Africa between 5,000 and 7,000 feet. It is largely, but not entirely, restricted to the drier areas on the northern and eastern slopes of the Ruwenzori chain, Mounts Elgon and Kenya, the Cherangani, Nyambeni and Chyulu Hills in Kenya and Mount Meru and Kilimanjaro in northern Tanzania. It is usually epiphytic on trees in the more open areas of the montane forests and on the scattered trees in the nearby savannah and partially cleared areas.

In cultivation the plants look best if they can be allowed to assume their natural attitude leaning outward and upward away from some form of support. This means that they can be planted in a pot or basket of fibre with the stem upright and then turned and suspended sideways so that the leaves hang downwards. Alternatively the plants can be really securely tied to an upright piece of hard wood in a similar position; a little moss or fibre as padding helps to provide a damp atmosphere near the stem to stimulate the production of new roots.

Climatically the plants need to be kept rather dry in temperatures of 70° to 80°F during the daytime. A considerable drop in temperature and a consequent rise in relative humidity will be found to be beneficial at night. In nature the plants are often found in very windy places so fresh air is probably essential. Abundant water during the flowering and growing season followed by a drier resting period will correspond to the alternating rainy and dry seasons which the plant normally experiences.

CHAMAEANGIS ORIENTALIS PLATE 21

CIRRHOPETALUM UMBELLATUM (FORST.F.) HOOK. & ARN.

The scent or odour of orchid flowers is as varied and interesting as their diverse colours and structures. Nevertheless it is somewhat surprising, when one finds an orchid species for the first time and immediately recognises it by its distinctive flowers, to be slightly repulsed by its unpleasant scent. When we collected *Cirrhopetalum umbellatum* on the slopes of Mount Karthala, on the island of Grande Comore, we noticed an odour of dead fish and were surprised to find that it emanated from the plants we were holding. The smell is presumably an attraction to the small flies which pollinate this flower.

Species of *Cirrhopetalum*, like their near relatives in the genus *Bulbophyllum* some of which are also carrion-scented, have a very delicately hinged lip as the chief distinguishing character of each flower. The lip is so carefully articulated that when an insect of a suitable size alights on it and crawls forward past the point of balance it is tipped head downwards into the flower so that its body touches the inner side of the column. As it frees itself it removes the pollinia and carries them away in such a position that they are deposited on the stigmatic surface of the next flower the insect visits.

The *Cirrhopetalum* species photographed here illustrates several of the features which distinguish the genus from *Bulbophyllum*. The flowers are arranged in an umbel-like inflorescence as opposed to a spike, and the dorsal sepal and petals have tufts of cilia at their tips. The lateral sepals vary in this species from 2·5–4 cm long and are twisted round with their lower surfaces uppermost and their outer edges joined for more than half their length. Several species of *Cirrhopetalum* have this character although in others the sepals are free. The long antenna on the dorsal sepal and the slightly shorter ones on the petals, together with the 5 mm stelidia projecting forward from the apex of the column, distinguish *C. umbellatum* from its allies as also does its very wide distribution. It was only quite recently that botanists at Kew realised that the specimens collected from Java southwards to New Guinea and across the Pacific Ocean as far as the islands of Tahiti and Fiji were botanically identical with those from Uganda and Tanzania on the African mainland and some from the nearby islands of the Indian Ocean. Related species occur on the Asiatic mainland but so far this particular species has not been collected there.

On Grande Comore the plants were found growing among epiphytic ferns with various species of *Bulbophyllum*, and in cultivation these are all being treated alike. They are potted in a fairly fine compost of chopped tree fern and leaf mould and kept in conditions which are intermediate as far as temperature is concerned, lightly shaded, and with a high humidity at night falling to about 50% during the day.

CIRRHOPETALUM UMBELLATUM PLATE 22

CYMBIDIELLA RHODOCHILA (ROLFE) ROLFE

Cymbidium is one of the most popular and well known of the orchid genera. Forty or so species and an increasingly large number of hybrids of this attractive genus have been grown in cultivation for about a hundred and fifty years.

Plants which have a rather similar mode of growth, and which produce heavy textured flowers that are *Cymbidium*-like, occur in Madagascar. For a long time it was thought that they too were representatives of this genus, although they are a long way from the main area of distribution which stretches from India to Japan and southwards to Australia.

By 1904 three species had been described in Madagascar and fourteen years later R. A. Rolfe proposed that these should be put into a separate genus – *Cymbidiella*, this name being a diminutive of *Cymbidium*. Some authorities now consider their true position should be nearer *Eulophiella* in another tribe of the orchid family but, whatever their correct taxonomic position is, the three species are as floriferous and attractive as many of the hybrid and highly prized Cymbidiums.

Each of the three species has quite distinct ecological requirements. One, *C. flabellata*, which has flowers about 5 cm across, is a terrestrial plant. It is confined to damp and often boggy places growing with sphagnum moss at the edges of some of the brackish lagoons on the east coast of the island and along rivers and streams inland. We have found it always in the shade of *Phillipia*, a heather-like bush, with its wiry rhizome a few inches above the damp soil.

C. humblotii produces a large, branched inflorescence with rather larger flowers than the above in the same shades of yellowish green but not spotted with red, and with the large lip outlined in a purplish black. This species is an epiphyte and is only to be found on the trunk of one of the larger palms which are so common in Madagascar. This could be said to be the second claim to fame for the palm, *Raphia ruffia*, the first being the quantity of raffia twine which it supplies for the use of gardeners all over the world.

C. rhodochila is even more specific in its choice of host plant. This orchid only grows with one of the 'Stag horn' ferns, *Platycerium madagascariense*, which is itself an epiphyte rather particular in its choice of habitat, commonly growing on *Albizia fastigiata*, a tall tree. It occurs only in a certain part of the forests on the eastern slopes of Madagascar at altitudes between 1,800 and 2,500 feet.

In cultivation this species grows best if it can be provided with plants of the *Platycerium* as companions rooted in osmunda or a tree fern fibre mixture. A rather large pot or basket is required to accommodate this *Cymbidiella*, especially as it shares with the other two species a strong dislike of having its roots disturbed, taking several years to recover after repotting before it will flower again unless it is transferred very carefully. It requires rather warmer conditions at night than *Cymbidium* plants but has similar daytime temperature and light requirements.

Once established *C. rhodochila* flowers annually. One or more racemes are produced from the base of the pseudobulbs bearing twenty or more flowers, opening in succession, each 10 cm in diameter. The flowers are long lasting and all the parts are very heavy in texture. The variety of colours in the four lobed lip is particularly striking and hybridisation of this plant with the other Cymbidiellas or related genera such as *Eulophiella*, offers some tantalising possibilities. Several breeders have already attempted to produce hybrids using *Cymbidium* as the other parent but so far without success.

CYMBIDIELLA RHODOCHILA PLATE 23

CYNORKIS FASTIGIATA THOUARS
C. PURPURASCENS THOUARS

PLATE 24A
PLATE 24B

The orchid species which first became known to European civilisation were those in Asia Minor and the eastern Mediterranean region which were studied by the early Greek philosophers and botanists. These were all terrestrial species with underground storage organs which made them capable of withstanding the long dry season. Without the use of lenses for detailed study the structure of the flowers was of small interest to the men who collected orchid plants at this time, but the roots, with their concentrated store of carbohydrates and alkaloids, were avidly gathered and studied.

Theophrastus used the Greek term '*orchis*', meaning testis, to describe several Mediterranean species of the *Orchidaceae* which had a pair of bulbous underground tubers each resembling this organ in shape. The name was perpetuated by Dioscorides three hundred years later and by the writers of several herbals during the Middle Ages. They believed that such plants, if eaten, must have some connection with improved fertility and virility in man and concocted recipes for their use accordingly. The name was retained by Linnaeus in his *Species Plantarum* for a genus of north temperate orchids which is now thought to number about thirty five species distributed from Madeira, across Europe, to southwest China.

Cynosorchis is also a name used by Dioscorides in the first century for an orchid which had two underground tubers which were thought to resemble a dog's testes in shape. This name was adopted in 1822 by Aubert A. du Petit Thouars for a genus largely confined to tropical and southern Africa and the islands of the Malagasy region, many species of which have this type of root. The shorter name, *Cynorkis*, which Thouars also proposed, is now the accepted one and about one hundred and twenty five species are known today from this area.

Photographs of two species of *Cynorkis* which are found in Madagascar and the Mascarene Islands are shown on the opposite page. *C. fastigiata* is also common in the Comoro Islands, further west, where this photograph was taken, and in the Seychelles further north. These two species illustrate several characters of the genus, in particular bright coloration, the small petals cohering to the edge of the dorsal sepal to form a hood over the column, and the pin-shaped glands on the outer surface of the sepals and on the pedicellate ovary. In addition the lobed lip is a brighter colour than the other floral parts and much larger in size, reaching a length of 1·5 cm in *C. fastigiata* and 2·5 cm in *C. purpurascens.*

Ecologically these two species are of interest in that they occur throughout their range as both terrestrials and epiphytes. In shady situations they flourish equally well on moss-covered rocks or in crevices between such rocks and on the damp moss-covered trunks of trees, often near watercourses. In humid areas the basal leaves appear before or with the flowers and in drier places usually only after the flowers have died. They are found from sea level to nearly 5,000 feet over a wide area, although *C. purpurascens* is much less frequently met with than *C. fastigiata.*

In cultivation these plants have succeeded in a very well-drained compost of leaf mould and loam mixed with an equal quantity of small chunks of lava cinders. Presumably other porous rocks or charcoal would do equally well but as the rocks among which the plants of *C. fastigiata* were collected were volcanic in origin it was decided to try the lava first. A resting period is necessary after flowering, as soon as the leaves begin to wither and die back; this corresponds to the dry season which the plants would endure in nature. The pots can be allowed to dry out almost completely provided they are kept in a shady and humid atmosphere so that the tubers do not become desiccated. They should be watered again heavily after three to four months or as soon as new shoots appear above the surface of the compost. Applications of weak liquid manure are probably beneficial during the growing season.

CYRTORCHIS ARCUATA (LINDLEY) SCHLECHTER SSP. VARIABILIS SUMMERHAYES

Cyrtorchis is an angraecoid genus restricted to Africa whose fifteen members all have elongated stems bearing two rows of thickened, fleshy or leathery leaves. The white flowers of this genus are easily distinguished from all other angraecoid orchids by their regular shape. The lanceolate sepals are only slightly larger than the petals and lip of similar form. All these parts are more or less recurved towards the pedicel and the lip only differs from the others in that it is prolonged at its base into the gradually tapering spur. All the species are strongly fragrant, particularly at night, and all have fairly large flowers; in the smallest flowers the sepals are about 1 cm long and in the largest up to 6 cm.

Cyrtorchis arcuata is the type species on which the genus was founded by Schlechter in 1914. Previously it had been described by Lindley under the name *Angraecum arcuatum*, and then transferred by Dr. H. G. Reichenbach to the genus *Listrostachys*; the latter is now a very small genus although at one time it was thought to contain a large number of angraecoid species which are now distributed among other genera. *Cyrtorchis arcuata* is widely distributed throughout the whole of Africa south of the tropic of Cancer, and is now recognised in several distinct forms in different regions; some of these had been thought to be species in their own right before large numbers of plants had been studied. The type subspecies is restricted to South Africa and parts of Mozambique and is separated by more than a thousand miles from the subspecies illustrated here. *C. arcuata* ssp. *variabilis* has the widest distribution, is the most variable of the five subspecies which have been described to date, and is a very common epiphyte throughout most of its range. It has been recorded from most of the countries across tropical Africa from Liberia in the west to Kenya in the east and as far south as the northern parts of Zambia and Malawi. It occurs in a variety of habitats, usually in rather thin shade, from sea level up to 8,000 feet, the highland forms usually having rather larger flowers than those from a lower level. The flowers are among the largest in the genus with sepals up to 4 cm in length and the spur in the largest forms reaching 6 cm. Another characteristic feature is the rather large bracts, one of which supports each flower; the bracts are pale green when the flowers first open but soon become brown or blackish. The flowers last for two or three weeks, provided they remain unpollinated, and as they fade to a dull orange over a period of a further week or two they still retain their scent.

A vegetative character which one observes when collecting this and other species of *Cyrtorchis* is the great length of the aerial roots, often extending for several yards over the surface of the tree in which the plant is growing. In Nairobi several plants have behaved similarly when fastened on to the smooth branches of a frangipani tree, the roots penetrating the soil six feet below in some cases. When the plants are grown in a more conventional way this feature must still be borne in mind, and if too many of the roots are removed or damaged when the plants are transplanted it may take several years for the plant to re-establish itself well enough to flower. Plants may be grown in fibre or bark of various kinds in pots but they will then surround the pot with their thick roots and often put out stragglers into the air or on to the bench as well. They grow in a tidier fashion in large baskets or securely attached to a piece of hardwood supported in a pot. Once they are well established they flower frequently, producing three or more spikes of five to ten flowers each from the axils of the leaves. Climatically this is a very adaptable species and in cultivation the plants can adjust to almost any semi-shaded conditions within a year or two provided the roots have plenty of space.

CYRTORCHIS ARCUATA SSP. VARIABILIS PLATE 25

EULOPHIA PETERSII (RCHB.F.) RCHB.F.

The habitats chosen by *Eulophia petersii* are almost the exact opposite of those preferred by the species on the following page. Throughout its range it seems to be able to withstand an astonishing degree of aridity and it flourishes in several areas which regularly receive less than twenty inches of rain per year. We have collected plants from Kenya's dry Northern Frontier Province, from the desiccated Olduvai gorge in northern Tanzania, and from sand dunes within a few yards of the high tide line along the Kenya coast. Although the rainfall is higher in this last area the plants probably experience conditions of physiological drought in the salty atmosphere there, and are exposed to drying winds and very strong sun. Elsewhere the species is widely distributed throughout the hotter parts of eastern Africa from Eritrea in the north to the northern parts of Natal and the Transvaal in South Africa.

The plants form quite large clumps among rocks and granite outcrops and in rather acid, sandy soils. They are frequently exposed to the wind and weather but are sometimes sheltered by tussocks of grass or low, woody vegetation. The very distinctive pseudobulbs could hardly be mistaken for any other plant, being entirely above the surface of the ground, usually yellowish and strongly ribbed, up to 30 cm tall and 3–4 cm in diameter, or sometimes much shorter and fatter. Two to five tough, leathery leaves arise from the apex of the pseudobulb. They are up to 40 cm long, have rough, scabrous margins, and are usually thickened so that at first sight they resemble the leaves of an aloe rather than an orchid. To add to this resemblance is the fact that this species mainly occurs in areas where one would expect to find aloes rather than orchids.

Other features which distinguish this species from most other Eulophias emerge when one finds it in flower. The inflorescence, arising from the base of a mature pseudobulb, is very tall and stout, frequently exceeds 2 m in height, and is often branched. The flowers are very widely spaced on the inflorescence but as it is so large there are dozens or occasionally over a hundred of them. The combination of green and purple tones of the ovary, sepals and petals shows up in the photograph and contrasts well with the white, pink-tinged lip with its rows of crests. Although the flowers of different plants are rather variable in size and coloration the sepals, and sometimes the petals, are often curled backwards at their tips, or circinnate, which is a further distinguishing characteristic of this species.

Its environmental requirements are so unusual compared with those of most other orchids that we have found it is best to treat this species as an aloe and plant it in a rock garden. In a dry year in Nairobi, for instance, it flowers under these conditions, but this is a marginal habitat for a species which does not seem to appreciate more than forty inches of rain per annum. In addition it may be best to be deliberately harsh to it and to provide it with little shelter, plenty of sun and a poor substratum; the plants might do better in pans of very sandy soil in the dry atmosphere of a succulent house rather than out of doors.

EULOPHIA PETERSII PLATE 26

EULOPHIA PORPHYROGLOSSA (RCHB.F.) BOLUS

Driving along the raised roads through the swamps of central Uganda there is much more of interest to be seen than is at first apparent. Initially one's attention is held by the enormous fronds of papyrus sedge. Each is three or four metres high and its feathery umbellate inflorescence, 30–40 cm in diameter, is constantly moving in the breeze. One feels as if one is moving along through a greyish-green undulating sea whose surface is just about at eye level. But sooner or later, depending on how fast one is travelling and how keen one's eyesight is, it becomes apparent that the flora of the swamp is far from uniform.

Along the edges, between the swamp and the road, and between the swamp proper and patches of open water many other flowering plants are to be seen, most of them with very conspicuous and interesting flowers. In the open stretches there are sheets of waterlilies with bright blue flowers sprinkled among their flat leaves. In the slightly drier places at the roadsides several yellow composites and a mixture of feathery grasses predominate. One of the giants among the orchids of continental Africa, *Eulophia porphyroglossa*, occurs here quite commonly and its bright mauve colouring makes it stand out against the greeny grey background of the papyrus whose height it almost equals.

In Kenya there are no comparable areas of open swamp in a highly humid climate. However, this *Eulophia* is found in the small areas of swamp which do exist, usually at a higher altitude than in Uganda, in situations where it is partly shaded by trees for at least some of the day. It also seems to have adapted itself well to a rather different habitat. On several occasions in different parts of the country we have found it growing right on the bank, or even in the edges of cool highland rivers, usually over 6,000 feet altitude. Although for much of the year the humidity is much lower than it would be in the centre of a swamp the temperatures to be endured are also lower and the plant has more shade.

Eulophia porphyroglossa puts up a new cluster of leaves every year. Each leaf is lanceolate, 100–130 cm long, and 10–15 cm wide with a strongly ribbed and plicate surface. During each season's growth a new potato-shaped tuber is formed at the base of the leaves. For the size of the plant these are rather small but they occur in long chains, each joined to that of the preceding growing season. The inflorescence arises from the developing tuber near the cluster of leaves. In Kenya we have seen plants in flower in May and October; both times coinciding with a rainy season. In Uganda, in the more equable climate, the flowering season is not so limited, and in different parts plants can be found in flower for most of the year.

Whenever they are found the 2–3 m spikes of flowers are a striking sight. The peduncle is erect and as thick as a walking stick; it is dark reddish purple in colour and always at least twice as long as the leaves; a fairly dense raceme of blooms occupies its terminal quarter. Each flower is 5–7 cm long and the rich mauve of the paired petals contrasts subtly with the bronze purple of the reflected sepals and ovary. The lip is basically pink or mauve with white or yellow frilled crests on its elongated central lobe, and the mauve-streaked, green lateral lobes provide a further contrast.

In cultivation this species is easy to grow in a rather large non-porous container such as a plastic tub or metal drum. A fairly rich loam which is always kept moist has proved a suitable medium. Plants also do well in some gardens in Nairobi in lightly shaded conditions at the edge of ornamental streams or ponds.

EULOPHIA PORPHYROGLOSSA PLATE 27

EULOPHIA QUARTINIANA A. RICH.

Part of a raceme of *Eulophia quartiniana* is shown in the photograph on the opposite page. The scape was about 40 cm high and bore twelve flowers, each of which was 6·5 cm long from the tip of the erect dorsal sepal to the front edge of the lip. Floristically it is one of the most desirable of this large genus, and fortunately it is one of the few which are easy to grow in cultivation.

It is a species with an interesting history. It was originally described by the French botanist, A. Richard, in 1851 from a specimen collected in Ethiopia by another Frenchman, Richard Quartin-Dillon, after whom it was named. Other botanists, however, thought that it was conspecific with a similar species, *E. guineensis*, which had been described from West Africa much earlier by John Lindley. The Belgian botanist Alfred Cogniaux gave yet another name to this species in 1895 when he described *E. congoensis* at its first recorded flowering in Europe. The botanical and ecological differences between the species have now been sorted out by Mr. V. S. Summerhayes[1] and it seems likely that most of the plants found in Kenya and other savannah regions of eastern Africa are in fact *E. quartiniana*, the name *E. congoensis* being a synonym of this species.

Vegetatively *E. quartiniana* and *E. guineensis* look very alike when dormant or in full leafy growth, and they can be equally easily grown in cultivation. They require rather different treatment, however; *E. guineensis* is a species of the forest floor, recorded from Gambia through similar habitats in west Africa eastwards to Uganda, and requiring considerable shade, moisture and a fairly rich compost; *E. quartiniana* is usually found growing amongst rocks, in only lightly shaded conditions, in the savannah areas of Nigeria and the eastern half of Africa as far south as the Katanga province of the Congo, and needs a very well-drained, gritty compost. Both species need a rest from watering in cooler temperatures after their growth is complete and appreciate repotting in fresh compost when new growth begins. Their main growing season in tropical Africa is in the period November to March with continued growth until July or August followed by a distinct rest. These conditions should be easy to copy in north temperate regions where a cool rest can easily be given during the winter.

The conical pseudobulbs of *E. quartiniana* are rather small, being 3–4 cm high, and sometimes half of this is buried in the fallen leaf litter and rocks among which the plant grows. The pseudobulbs bear the circular scars of old leaf bases on their dark green surface and are anchored to the substratum by numerous thick white roots. At the approach of the rainy season in November the conspicuous flowering scape appears, either with, or shortly before, the growth of a new leafy shoot. The flowers last for several weeks and have a very pleasant, delicate scent. After they have faded the leaves continue to develop and when mature there are four or five of them each 30–40 cm long and 8–12 cm wide with a petiole taking up half this length. The new pseudobulb develops within the bases of these leaves.

[1]Curtis's Botanical Magazine, **171**, tab. 267 (1956).

EULOPHIA QUARTINIANA PLATE 28

EURYCHONE ROTHSCHILDIANA (O'BRIEN) SCHLECHTER

Among the many African orchids described by Dr. Rudolf Schlechter are two members, the only two, of the genus *Eurychone*. He founded the genus in 1918 to accommodate these attractive and unusual plants with their very broad, funnel-shaped lip (Greek, *eurys*, broad; *chone*, funnel).

Each flower in the short axillary raceme of *E. rothschildiana* is approximately 6 cm in diameter and the lip itself is just over 2·5 cm across. All the floral parts are basically pale green, with darker green veins and a narrow white border, although some forms occur which are almost entirely white. The lip has a large blotch in its throat which varies from chestnut brown to purplish or almost black in different plants, and in front of this is a brilliant green area contrasting well with the margin of white which is distinctly frilled on its outer edge. It is when viewed from the side that this lip most clearly resembles a funnel as it narrows rapidly behind the dark coloration and is projected beyond a distinct constriction into an almost utriculate spur about 1·5 cm long. The central column is fairly large and in the flower on the right-hand side of the photograph the anther cap has been damaged and partially removed, exposing the bright yellow pollinia which are borne on a single caudicle.

This *Eurychone* has a very delicate scent as also does the smaller species, *E. galeandrae*, which has white or pale pink flowers with wine red streaks on the inside of the lip. Both are plants of the tropical rain forests of Africa; the one illustrated here has been recorded from Sierra Leone, Ivory Coast and across to Uganda, and the other has so far been collected only in the Congo and Gabon. Plants which we collected near Kampala and Entebbe, and also in the Ituri Forest near the border of the Congo with Uganda, were thriving on the thin trunks of young trees in a ground-water forest, and also on the woody lianes or creepers which thread themselves among the trees. Temperatures and humidity were both high and the plants were all in the relatively still atmosphere near the ground and up to twenty five feet above it.

The plants resemble those of the genus *Phalaenopsis* in size and shape although the two to three pairs of dark green leaves are thin and leathery rather than succulent. In cultivation *E. rothschildiana* does well under the same conditions of warmth, humidity and shade as are usually recommended for growing *Phalaenopsis*. In pots or baskets of suitable fibrous composts the plants flower regularly, producing two or more racemes of up to seven flowers from the axils of the leaves at least once a year. The spikes are somewhat shy and need to be eased away from the shelter of the leaves so that the flowers can open properly and be fully appreciated. Like many other angraecoids, however, the plants grow better when firmly attached to a hardwood log suspended in the greenhouse. They grow away from the log at a right angle as they grow on their host tree or liane in the forest. The short sprays of flowers hang down below the leaves and are much more accessible to view, or for an insect to pollinate, than when the plant is grown in an upright position in a pot.

EURYCHONE ROTHSCHILDIANA PLATE 29

GRAMMANGIS ELLISII (LINDLEY) RCHB.F.

Botanical exploration began almost two centuries earlier in Madagascar than in the countries of East Africa and over a hundred years earlier than in the neighbouring mainland country of Mozambique. Plant specimens still exist in the *Museum national d'Histoire naturelle*, in Paris, which were collected by Etienne de Flacourt during his period as an administrator in the island from 1648 to 1655. When he was on his way back to the island in 1660 his small flotilla of boats was attacked and destroyed by pirates in the Mediterranean.

After this it was almost a hundred years before any more important botanical collections were made in Madagascar. Then, from the second half of the eighteenth century onwards, many collections, large and small, were made by botanists and other visitors.

The naturalist Aubert A. du Petit Thouars was probably the first visitor to the island who was particularly interested in orchids, and from 1793 to 1802 he collected in Mauritius, Madagascar and Réunion. He took all his specimens back to Paris and worked on many of them there himself publishing diagnoses of the large genera *Bulbophyllum* and *Cynorkis* and descriptions of many species from the area.

As in other parts of Africa great contributions to our present-day knowledge of the Madagascan flora were made by people who collected plants as a hobby. Doctors and naval surgeons, hydrographers and surveyors, administrators and consular staff, and settlers and missionaries of many different European nationalities collected and pressed plants in the little spare time that was available to them when their duties for the day were over. One of the English missionaries, Rev. W. Ellis, collected in the central parts of the island from 1856 to 1865 and the *Grammangis* pictured here commemorates his work, as also do several other Madagascan orchids.

Grammangis ellisii is a large epiphyte of the eastern forests of Madagascar. It usually flowers at the same time as it produces a new growth, in January, the inflorescence and new vegetative shoot appearing together or from opposite sides of the base of a mature pseudobulb.

The developing pseudobulb is enclosed by bracts when young but these are deciduous. By the time it is mature the pseudobulb is usually bare, distinctly four sided, and up to 10 cm long with a tuft of leathery or fleshy leaves at its apex. Each leaf is 15–40 cm long and parallel-sided, being 1·5–4 cm wide and folded towards the base.

The inflorescence is a simple raceme, usually longer than the leaves, and forms an attractively arching spray of fifteen to forty flowers. These are usually well spaced on the strong thick peduncle and up to 8 cm in diameter. The coloration varies from one plant to another but the broad sepals with their wavy margins always have a basic ground colour of yellow and light or heavy chestnut brown markings. The petals are smaller and paler in colour and project forward in the shape of a hood over the column. They are joined at the base to the lateral sepals which, enclosing the foot of the column, form a distinct mentum in place of a spur. All the perianth parts are thick and rather waxy and the flowers are long lasting.

Once established in cultivation this species is easy to grow in intermediate conditions of temperature and humidity. It does not like being disturbed, however, and even small plants should be potted with plenty of space for growth in osmunda or tree fern fibre rather tightly packed. Conditions of semi-shade suit this species best during the growing period and fairly frequent applications of a liquid fertiliser seem to be beneficial. When the growth of a new pseudobulb is complete the plant should be rested for a month or two and then placed in a brighter light to encourage new growth and flower production.

GRAMMANGIS ELLISII PLATE 30

JUMELLEA COMORENSIS (RCHB.F.) SCHLECHTER

Schlechter established the genus *Jumellea* in 1914 and since then some forty four species have been described from Madagascar and the Comoro Islands, eleven from the Mascarene Islands, and one trom the eastern part of continental Africa. The genus was so named in honour of Professor H. L. Jumelle, a French botanist from Marseilles, who worked on the flora of Madagascar and several French colonies in Africa during the last years of the nineteenth century and the beginning of the present.

A distinguishing character of the genus is that the inflorescence usually bears a single white flower, or, rather rarely, two flowers. Certain species of *Angraecum* also bear only one white flower but those of *Jumellea* are rather easily recognised as such by their curious structure. The dorsal sepal is always carried more or less upright, or bent slightly backwards, while all the other floral parts project downwards and forwards. An explanation for this posture is found when the flower is dissected.

At the base of the column there are two short parallel arms which extend forwards across either side of the mouth of the spur and are joined to the surface of the lateral sepals. They reach as far as the edge of the petals, to which they are also joined, so that these perianth parts are stiffly held in the downward position with their outer surface uppermost. In this way they form a frame on either side of the lip, balancing the upright dorsal sepal, and giving a distinctly bilabiate appearance to the flower.

Jumellea comorensis is a rather straggling plant which we collected on the island of Grande Comore on the trunks of rather roughly barked trees at about 1,500 feet. Flowers were borne singly and occasionally in pairs on several branches of the plant in January and were clearly seen among the dark green glossy leaves. The narrow sepals and petals were each 2 cm long and 4 mm wide. The lip was the same length and up to 7 mm wide at its widest part, while the filiform spur was 12 cm long and narrowed gradually towards its tip.

Examination of the natural habitat for specific requirements indicated that this plant would probably need fairly bright light and cool to intermediate temperatures with high humidity in cultivation. In the forest in which it was collected the annual rainfall sometimes exceeds two hundred inches but there are often strong winds and the air is always fresh. The plants were usually pendent with their basal roots growing among epiphytic mosses and ferns and decaying plant detritus; other roots were creeping over the bark of the host tree and holding the plant securely in position.

Several plants of *Jumellea* species collected on Grande Comore were potted in a mixture of bark, shredded tree fern fibre, and small lumps of charcoal. They were sprayed with water three or four times a day and within a month new roots formed at the base of the plants and penetrated the potting mixture. Since then *J. comorensis* has been in flower almost continuously and both the small plants which are kept upright on the bench, and the plants with longer growth whose pots are suspended on their sides from the roof of an intermediate greenhouse, are an interesting and attractive addition to the collection.

JUMELLEA COMORENSIS PLATE 31

LIPARIS NEGLECTA SCHLECHTER

Cultivation of orchid plants which are annually or seasonally deciduous is less easy in tropical countries than in temperate ones. In the latter there is a natural period of lower temperatures which reminds one to refrain from watering the plants at that time. They can die back and have a complete rest in almost dry compost before shooting into growth again with the rising warmth of spring. In the tropics this regime exists to some extent in nature in the alternating wet and dry seasons. Many plants seem to be able to withstand very dry periods, of variable duration, in the dormant stage, completing growth and flowering in a very short time during the rains. Trying to adapt plants which are grown within the confines of a small pot or basket to this system is not easy, however, even when the plants are kept out of doors and exposed to the natural climate.

The species of *Ancistrochilus*, *Bonatea*, *Brachycorythis* and *Liparis* included in this book are all plants which have this pattern of growth. Although it is sometimes hard to stop watering them during the hot dry season this is essential as soon as they begin to die back; when successfully grown both the plants and their flowers are most rewarding. Several plants of these genera have been disappointing, however. They flower well for two or three seasons but do not survive longer. Perhaps they are naturally rather short-lived, or it may be that the technique of growing them has not yet been perfected.

Liparis neglecta is interesting in that it occurs in many parts of eastern Africa as a terrestrial, growing in the leaf litter of the forest floor. In other parts, notably in Kenya, it is found at altitudes of over 7,000 feet as an epiphyte on dead or dying trunks of the tree fern, *Cyathea manniana*. The plant shown in this photograph is grown in a mixture of old tree fern fibre, crushed dead leaves and sharp sand; it was watered daily during the short period of growth and flowering, and allowed to dry out as soon as the leaves began to turn yellow. A prolonged dry period under the bench then followed, with no watering until there were signs of renewed growth several months later.

This is one of the smaller members of a genus with about two hundred and fifty species of almost cosmopolitan distribution. Some species of *Liparis* have been found in almost every part of the world except New Zealand, varying in size from a very few to 50 cm high. *L. neglecta* has conical pseudobulbs 2–3 cm high standing close together on a basal rhizome. Each new growth is about 15 cm high with two pairs of broadly ovate, plicate leaves which are a pale green and have distinctly sinuous margins. Up to ten flowers are produced on each stem, usually opening in succession and fading to a dull orange or ochre colour after one to two weeks.

In all members of the genus the structure of the flower is unusual in that the petals are often very much reduced, sometimes being almost thread-like, and curving down and forwards. The lateral sepals are correspondingly enlarged and curve forwards immediately below the lip. In some species these have fused into a synsepalum and their shape and disposition almost present the appearance of a double lip. The latter is kidney-shaped in *L. neglecta* and has a single bright green, shiny patch, sometimes restricted to a median groove, in its centre. The green column curves forward immediately in front of the narrow dorsal sepal.

LIPARIS NEGLECTA PLATE 32

MICROCOELIA GUYONIANA (RCHB.F.) SUMMERHAYES

Aerial roots of epiphytic orchids have a number of functions in addition to the absorption of water and nutrients which is a common purpose of roots throughout the plant kingdom. The one which springs quickly to mind is that of anchoring the orchid plant firmly to its host where it will survive for many years despite high winds and storms, and even the depredations of orchid hunters. Anyone who has tried to detach a *Cyrtorchis* or *Microcoelia*, or indeed plants of any of the genera in the subtribe *Aërangidinae*, without damaging the roots in the process will know how very efficiently this function is carried out. Some genera, such as *Vanilla*, produce a 'fur' of root hairs on the surface of the root which is next to the host plant but others seem to cling on entirely by their 'skin'.

Absorption is carried out effectively by epiphytes during every storm of rain or night of heavy dew. The internal structure of an aerial root is similar to that of all roots of monocotyledonous plants except that it is entirely surrounded by a layer of air-filled cells, the velamen. This layer acts like a sponge for any moisture that is available. Many local epiphytes in cultivation in Kenya grow best out of doors, either in baskets, attached to pieces of bark, or transplanted to suitable trees in one's garden. Dew is often heavy, especially when the temperature drops by 40°F at night during the long, hot, dry season, and the plants are subjected to more or less the same conditions of wetting and drying out that they would experience in the surrounding countryside. Although the amount of nutrient material available in dew and rain-water is small epiphytic orchids are very slow growing and do in fact obtain all that they require in this way.

In many species photosynthesis takes place in the roots as well as in the leaves, chlorophyll being present in the cortical layer of the root. Some genera, such as *Microcoelia*, are almost completely leafless. Small brown scales, which are morphologically leaves, exist on the short stout stems, but these leaves appear to be vestigial and the entire photosynthesis of the plant is taken over by the roots. It is somewhat surprising, in this case, that the roots are not green in colour, but the chlorophyll-containing layer is below the velamen, or air-containing layer, which is thus seen to have a protective function. Its white colour and the many layers of cells probably also insulate the inner parts of the root from radiation and consequent overheating.

Most species of *Microcoelia* are difficult to find in the wild, being greyish-white like the bark of the quite small bushes and branches of trees on which they grow. Large specimens, such as that shown here, can be mistaken for birds' nests although they are more easily recognised when in flower. Plants are best collected with the branch on which they are growing as re-establishing them on another tree or piece of bark is by no means easy, and they will not grow in pots or baskets of the conventional fibre.

There are some twenty seven species of *Microcoelia* in tropical and south Africa and Madagascar, one small member of the widespread genus *Taeniophyllum* in Ghana and the Congo and two monotypic allied genera in west Africa. Leafless orchids also occur in other parts of the world and this type of plant growth perhaps reaches its greatest development in New Guinea where there are many species of *Taeniophyllum*. In some of these the aerial roots have become so modified for their photosynthetic function that they have assumed a leaf-like appearance. The genus *Polyrrhiza*, which has three species in the West Indies and one in Florida, is perhaps the best known of the leafless orchids. In the Florida species the greenish yellow flowers are sometimes as much as 12 cm long. Most of the other leafless orchids, including *Microcoelia*, have rather small flowers. They make up for their size in numbers, however, and a large plant flowering well is an attractive and unusual addition to an orchid collection.

MICROCOELIA GUYONIANA PLATE 33

NEOBATHIEA FILICORNU SCHLECHTER

Collecting non-flowering orchids in remote areas is an act of hope rather than wisdom especially when the plants are small and one is a long way from home. During a holiday in the Comoro Islands we spent eight days collecting orchids in various habitats on the island of Grande Comore and although it was at the beginning of the rainy season, when many orchids are coming into flower, we came across many non-flowering plants which we were quite unable to identify. The larger ones were fairly distinctive and with notes of the species we expected to find it was not difficult to name them, at least to the genus, without waiting for the flowers. But after one successful hunt along the bed and banks of an intermittent stream we found we had amassed a large bag of unknowns, each with a very short stem and three or four elliptic green leaves of varying size. When sorting them out later on it was obvious that the plants represented at least five or six species, probably more, although some might turn out to be merely young plants of species that we had already collected and named.

The first to flower in cultivation, some three months later, was a little stemless plant with three leaves which were only 4 cm long and 1 cm wide, narrowing towards the base and rounded at the tip. The inflorescence, bearing only one flower, arose below the leaves, and its increasing growth was watched avidly for more than two months while speculation as to its identity increased. The peduncle was only 3 cm long but the dark green bud grew daily and the small plastic pot in which the plant was established had to be perched on top of another to accommodate the increasing length of the spur which eventually attained 14 cm. The flower opened a pale green and quickly turned to the bright shade of yellow at which it was photographed. All the parts of the flower were thick and rather waxy in texture, the sepals about 14 mm long, the petals slightly shorter and the pointed lip more than 2 cm long.

The description and a sketch of *Neobathiea filicornu* in the standard work on the *Orchidaceae* of Madagascar[1] fitted our plant exactly, except that the author states that the flower of this species is pure white and that the species is confined to Madagascar. Mr. P. F. Hunt was able to confirm this identification when the flower was sent to Kew, and to inform us that this species is relatively common on the island of Grande Comore. The question of the yellow colour remains something of a mystery; it might be due to our cultural conditions, particularly the amount of shade, so it will be interesting to see what happens when the plant flowers again.

Unfortunately only four of our small plants appear to be the same as this exciting species and no more of them have flowered so far. Two other species have also flowered, a *Chamaeangis* and an *Angraecopsis*, both unidentified as yet, and we are eagerly watching the other small plants. They are all similarly potted in tree fern fibre, and kept in intermediate conditions of temperature and humidity and semi-shade.

[1]Perrier de la Bathie, H. (1941) in Flore de Madagascar edited by H. Humbert. 49e Famille, Orchidées. Tome II.

NEOBATHIEA FILICORNU PLATE 34

OBERONIA DISTICHA (LAM.) SCHLECHTER

Perhaps John Lindley was in one of his whimsical moods when he adopted the name of Oberon, the legendary king of the fairies, for a genus of miniature orchids which have minute, but very numerous, fairy-like flowers. He wrote 'As Oberon, that little King of the Dryads, prince of the northern hob-goblins, rides about on the branches of trees, hiding his many formed countenance among the leaves, so our little herbs, not less changeable in form, lurk in the forests of India and ride triumphantly in their leafy chariot'. He was undoubtedly referring to the character of the lip in this genus, which is minute but extremely variable and is often the distinguishing character between vegetatively similar species.

Oberonia is a large genus of over three hundred species, whose main centre of distribution is in the tropical parts of Asia and the islands of the western Pacific, with one species known at present from Africa and the Malagasy region. All members of the genus are easy to recognise by their very flattened shape, looking as though they had spent their life in a plant press or other confined space. The stems are produced successively, close together on a scarcely discernible rhizome, and are covered by the alternate fleshy leaves. The latter are so much flattened that they have no upper surface except for a minute amount at the sheathing base, and in this they resemble *Angraecum leonis* (page 45) which is also an epiphyte in areas of high rainfall. Another distinctive character of *Oberonia* is the terminal inflorescence which has been variously described as resembling the tail of a rat and a fox. It is made up of hundreds of dainty flowers arranged in whorls of four or five and it continues to grow at the base after the flowers at the tip of the spike have begun to open, or after the central flowers of the spike have begun to open in some Asian species. The entire inflorescence lasts for about two months.

The flowers of *O. disticha* are larger than those of some other species of this genus being individually almost 2 mm long. Their delightful texture and shape can be seen easily with a hand lens, and the fiddle-shaped lip with its fringed tip is particularly intriguing. The inflorescence is frequently as long as, or longer than, the leaves so that the total length of an individual plant may be as much as 20 or 30 cm although it is usually less.

This species is very common in suitable areas of rain forest from the Guinea coast of west Africa to the warm wet slopes of the mountains of the Comoro Islands, Madagascar, and as far eastwards as the island of Rodriguez. It is always found in warm, humid conditions and in cultivation needs to be kept at 55°F or more at night, rising to 80–90°F during the day, with relative humidity rarely less than 60% if possible, and in semi-shade. The roots are very thin and wiry and the compost in which they are potted needs to be rather well drained. Finely chopped tree fern fibre mixed with small lumps of charcoal has suited the plants we collected on the island of Grande Comore which are shown in the photograph here. On their forest hosts the plants frequently grow in an upright position with their inflorescences curving over as in this photograph, and they were also found hanging straight down from the branches to which they were attached.

OBERONIA DISTICHA

PLATE 35

OEONIELLA POLYSTACHYS (THOUARS) SCHLECHTER

One of the distinctions of epiphytic African orchids is that so many of the species have white flowers. Although not large enough, except for a few of the Angraecums, to be much in demand for floristry work on their own, whole racemes of these flowers are sometimes required for corsages or the delicate tips of bridal bouquets. Species of *Aërangis* and *Rangaëris* are particularly popular but unfortunately most brides do not choose the rainy season, when many of these species flower, for their wedding day, and orchid growers are frequently asked for these fragile-looking flowers in the middle of the hot dry weather when they are quite unobtainable. A collection of plants of *Oeoniella polystachys* from different parts of its range in Madagascar and the Comoro Islands, on the other hand, would be ideal for this purpose, as it can be found in flower throughout the period September to May in different parts.

The flowers from different latitudes vary considerably in size as well as flowering season; those from Grande Comore are said to be the largest, up to 6 cm in diameter, but the ones photographed here from that island were not quite so large. The lip is the most distinctive part of the flower, being trumpet-shaped and three lobed at its tip. The pronounced median lobe is narrow and pointed while the rounded side lobes have frilly edges. The short greenish spur at the base of the lip provides the only patch of colour in the flower. This short spur makes the flowers very much easier for a florist to handle than those of *Rangaëris* or *Aërangis* so they are always very welcome when they can be provided.

Seven to twelve flowers are produced on each spike and there are usually two or three, and sometimes many more, spikes at each flowering. It was presumably from this feature that the specific name was first derived (Greek *polys*, many; *stachys*, spike) by Thouars in 1822 when he described this species as a member of the genus *Epidendrum*. Since that time it has appeared in no less than four other genera, but at least always with the same specific name, until it came to rest in one of Schlechter's new genera, *Oeoniella*, in 1918.

Part of the fleshy foliage can be seen in the photograph of a fairly robust plant of this species. It has a typically vandaceous type of growth and some of the plants we collected from a tree in Grande Comore were more than 50 cm in length. Huge clusters of upright stems surrounded the trunk and lower branches, growing parallel with the host plant, their aerial roots all curving round to one side to anchor themselves on to the bark of the tree. Planted in a polystyrene pot with tree fern fibre and charcoal these plants have continued to grow and flower without any setback following their removal and transportation. Many of the roots are somewhat difficult to accommodate in this sort of container, but like some of the strap-leaved Vandas with which they are grown it does not seem to matter if the roots protrude into the air in all directions. Presumably if they were provided with a stake they would grow on to it in the same way as they grow on the trees naturally. The plants were always found on trees with a thin canopy, most often on the trunks, in low-lying forests or on isolated trees near the shore in conditions of considerable warmth, high humidity, and fairly strong light.

OEONIELLA POLYSTACHYS PLATE 36

POLYSTACHYA BELLA SUMMERHAYES

More than two hundred species of *Polystachya* have been described from the tropical and subtropical parts of the world, most of them from the African continent. It is one of the genera in which the ovary is not twisted, or resupinated, so that the lip, which protrudes between the two more or less elongated lateral sepals, is uppermost in the flower. This is clearly shown in the photograph of *P. bella.*

This species has been known in cultivation for many years, both in Kenya, where it occurs naturally, and at Kew, where it has flowered many times since its introduction in 1934. After its discovery in the Kericho district of Kenya in 1929, however, it was confused with the Nigerian species, *P. obanensis* Rendle. It appeared under this name, both in the literature and in many orchid collections, until 1960 when it was recognised and described as a distinct species by Mr. V. S. Summerhayes. He states that the differences between the Kenya species, *P. bella*, and the Nigerian one are that the Kenya plant has a more dense inflorescence with much longer bracts, the individual flowers a deep golden colour instead of pale cream, the perianth parts much longer and narrower, and the lip distinctly fleshy. The inflorescence always arises from the apex of the developing pseudobulb and bears a maximum of thirty five flowers each about 1·5 cm long.

This species is one of those in the large section *Cultriformes*, which is characterised by the fact that the pseudobulb is always formed of one node and usually bears only a single leaf. There is great variety in the shape of this pseudobulb within the section, and that of *P. bella* is most unusual in being almost oval and laterally compressed, about 4 cm long and 2 cm wide. At first sight it looks very like some of those in the American genus *Oncidium*. The plant reaches 15 or 20 cm in height and the pseudobulbs stand very close together, upright, and with a rather dense mass of wiry roots at their base.

The plant is thus a very convenient shape and size to grow in a pot or basket on a mixture of bark, charcoal and osmunda or tree fern fibre. In quite a short time large and very floriferous plants can be established in warm moist conditions where the temperature does not fall below 55–60°F at night. An abundant supply of water is necessary during the growing and flowering season and after this the plant should be given a slight rest with only sufficient water being supplied to prevent shrivelling of the pseudobulbs.

POLYSTACHYA BELLA PLATE 37

POLYSTACHYA CULTRIFORMIS (THOUARS) LINDLEY EX SPRENG.

POLYSTACHYA CULTRIFORMIS VAR. AFRICANA — PLATE 38A

POLYSTACHYA CULTRIFORMIS VAR. AUTOGAMA — PLATE 38B

Slight variations in colour between individual plants of a particular species are as common among orchids as in other families of flowering plants. In horticulture, particularly on the commercial side, colour is a very important feature and many varietal names are given to plants on this basis. In botany, however, a colour difference has to be allied with a structural or at least a size or habitat difference before the plant is described as a distinct variety or a new species.

The two plants pictured here were both collected in the same small patch of forest at the southern end of the Aberdare mountains in Kenya. Several paler shades of pink were collected at the same time. In the Kericho area of Kenya, where this species is very common, collectors have claimed that they can distinguish as many as thirty different coloured forms. These range from white through cream to a deep, almost orange yellow, and in another series through pale shades of pink to a deep mauve. We have also seen one from this area which had flowers which were almost pale blue, although they still had some pinkish tinges.

This is a common orchid throughout highland areas of tropical Africa, occurring also in Madagascar, and the Comoro and Mascarene Islands. On the mainland the pink and mauve varieties are most common between 6,000 and 8,000 feet and Schlechter has separated these into the variety *africana*. The yellow flowered plants, which are found almost exclusively above 8,000 feet, he has placed in the variety *autogama*. A further difference is that in the yellow forms the rostellum is so shallow or even absent that self fertilisation of the flowers appears to be the rule. In Madagascar and the nearby islands the flowers are usually white and the plants are widely distributed from sea level to 6,000 feet. In the higher areas the flowering season is restricted to the summer months (December to March) but elsewhere plants can be found in flower all the year round, as on the African mainland.

From its wide range of habitat it is obvious that this is an orchid species with very catholic requirements as far as cultural conditions are concerned. Plants range from a few to 30 cm in height and grow well in pots or baskets on a mixture of osmunda or other fibre and leaf mould. The inflorescences are produced from the apex of the developing pseudobulb, as in *P. bella*, and like that species *P. cultriformis* requires plenty of water during the growing season and a rest after flowering.

POLYSTACHYA TAYLORIANA RENDLE

There is a small section of the genus *Polystachya*, the *Dendrobianthe*, in which the species have rather differently shaped flowers from all the others in the genus. The conical pseudobulbs bear several leaves when young but these are deciduous and the flowers are usually borne from the apex of the recently developed pseudobulbs after the leaves have fallen. They are open and flat and up to 2·5 cm in diameter. The colour varies considerably from white to pale rose or lilac and a variety of markings have been described on the lip. In shape and colour the flower reminds one strongly of some species of *Dendrobium*, so this section is well named.

Polystachya tayloriana also occupies a most unusual ecological position for an epiphytic orchid. Plants are found in the hot and arid regions in the east of the continent from Kenya to Rhodesia, and in the west there are records from Congo and Angola. Over this wide area of predominantly dry country they occur either as terrestrial or epilithic plants on rocky outcrops or, more commonly, as epiphytes on the old fibrous stems of dead or living *Vellozia* plants. *Vellozia* is a xerophyte akin to *Aloe*, and the genus has several species which develop hollow fibrous stems up to 2 m or more in height whose consistency very closely resembles osmunda fibre. They thus present an ideal surface for an epiphytic orchid, although they do not normally occur in what one would have thought was a suitable climate or situation. However the structure of the *Vellozia* stems and the enormous length of some of the roots of this *Polystachya* compensate for this. The stems of the host, being hollow, provide a moist microclimate for the roots, which penetrate inside, long after the brief rainy seasons are over. Furthermore these roots are extremely long and although the orchid plant may be situated several feet above the ground on the outside of the *Vellozia* plant its roots on the inside frequently reach and even penetrate the earth. The Vellozias themselves always grow among rocks where the rain water runoff is great and the amount of water available to plants growing in the soil is considerably greater than might be imagined. When growing as epiliths or terrestrials plants of this *Polystachya* are also always found in similar situations.

At different times a variety of names have been given to this orchid in different parts of its range. It seems likely, however, that there are in fact only two species with this curious habit: one figured here, and the other *P. zuluensis* L. Bol. which differs from *P. tayloriana* by its rhomboid lip and narrower curved petals. These two species are distinct also in the range they occupy. *P. zuluensis* has only been recorded in the north of Zululand and on the Lebombo plateau of South Africa and Swaziland, and some six hundred and fifty miles separate it from the southernmost record of *P. tayloriana* in Rhodesia.

This species grows best in strong light and in a low humidity provided that the roots are kept constantly moist during the brief growing season and allowed to dry off afterwards. Temperatures may be allowed to fall to 45°F at night and rise to 75 to 80°F during the day. Under these conditions a single stem often remains in flower for a year or more.

POLYSTACHYA TAYLORIANA PLATE 39

POLYSTACHYA VULCANICA KRAENZLIN

The Impenetrable Forest of western Uganda covers a volcanic area of extremely steep ridges and gorges. It is mainly composed of very tall trees, with few bushes and lianes except near streams and footpaths where more light can penetrate. For much of the time the very steep slopes are covered with slippery, decaying leaf mould and mud, which make it a very difficult forest to walk around in although the vegetation is not so dense as in many others in East Africa. The plants in it are not nearly so difficult of access as one is led to suppose by the name inflicted on the area.

The name arose when a boundary line between the countries to be administered by Belgium, Germany and Britain (Congo, Tanganyika and Uganda) was being surveyed. Not surprisingly, in an extensive, steep and rugged area which could only be forested in a tropical climate, the surveyors had a very difficult time. Cutting back the forest to provide tracks along which they could survey the terrain from one hilltop to another, or trying to hoist their instruments to the treetops to give a clear line of sight in uncut areas, proved equally formidable tasks. It must have been almost in desperation that the surveyors wrote across their maps of this area 'impenetrable forest'. As so often happens, when maps are redrawn far from their country of origin, this eventually became a proper name for the area concerned, and in some ways a fairly accurate one.

The greater part of this forest is cool by tropical standards, as it is situated between 6,000 and 9,000 feet. Botanically the area is of great interest; it has proved an attraction to most botanists visiting or working in Uganda and many orchids have been collected there. *Polystachya vulcanica* occurs there and also in the forests on the slopes of nearby volcanoes in the part of Congo north of Lake Kivu, and in Rwanda, on bare or mossy branches of forest trees. It is another of the species in the *Cultriformes* section of *Polstachya*, with one thickened linear leaf surmounting each slender, cane-like pseudobulb. The 3–5 cm pseudobulbs stand close together in a dense mat and from a little distance have the appearance of a tuft of grass. It is easy to grow in a mixture of fibre and leaf mould in semi-shade under cool, airy conditions. When a large plant is established the masses of pinkish mauve flowers 10–15 mm in diameter, with a distinct mauve lip and anther cap, are a most rewarding sight.

POLYSTACHYA VULCANICA PLATE 40

RANGAËRIS AMANIENSIS (KRAENZLIN) SUMMERHAYES

Judging by its wide distribution in the drier parts of east tropical Africa this orchid should be a very easy species to cultivate. It is usually the first indigenous orchid one meets on a visit to Nairobi: very large clumps of it are common on trees in the city and its environs. It is also one of the few epiphytes which thrives equally well in Kenya's dry Northern Frontier Province, the climatically similar Karamoja District of Uganda, the Rift Valley of Kenya, and the much more humid Usambara mountains of Tanzania. It is fairly catholic in its choice of habitat, doing just as well on isolated trees in some of the extensive grasslands in this part of the world as in the dry evergreen forests and thickets which are found at slightly higher altitudes. It has also been found growing among rocks in these localities. Unlike many other epiphytes its demands for moisture are apparently negligible or, at least, are easily met. Large plants survive and increase in some of the driest situations where the rainfall may be as little as fifteen inches per year and where even this is concentrated in a few heavy storms. Humidity is high during the actual rainstorms but for much of the year, and even during the rainy season, it falls to 30% or less by day rising at night when there is a heavy fall of dew. Temperature variations in the areas where this species grows are considerable, frequently reaching 90°F during the day and falling to 45°F at night or even lower. Its only idiosyncrasy seems to be with regard to light, as it cannot thrive in direct sunlight or flower in too much shade. It is equally tolerant of almost semi-desert and dry montane forest conditions and has been found throughout an altitude range of 3,000 to 8,000 feet. In fact it is a very easy plant to grow, provided it is not pampered, and large plants do well in baskets of bark or tree fern fibre, or on rafts of the same material.

Vegetatively this *Rangaëris* is a very distinctive plant with its two rows of short leathery leaves which are equally bilobed at their tips. The stems are long and often branching so that large clumps of the plant are found more often than single stems. The aerial roots arise below and among the leaves and are extremely thick, sometimes almost 1 cm in diameter, with a very wide chlorophyll-containing cortical layer.

The delicately poised sprays of flowers are always produced during the rainy season. They vary somewhat in size: some flowers are only 2 cm across but plants do occur in which they are over 5 cm wide. Those on the plant illustrated here were 3·5 cm across and 4 cm long with a 15 cm spur. The flowers are always pure white except the slightly curving spur in which there are often green or brown tints.

Six species of *Rangaëris* are now recognised in tropical Africa and one of them (*R. muscicola*) extends its distribution as far south as Natal. They were originally regarded as part of a section of the genus *Aërangis* (of which name the present genus is nearly an anagram) but the section was raised to generic rank by Mr. V. S. Summerhayes. The fact that each pollinium is supported by its own stipes is the chief character which now distinguishes it from the parent genus. This can easily be seen in *R. amaniensis* with a small hand lens when the anther cap is carefully removed from the column.

RANGAERIS AMANIENSIS PLATE 41

SPHYRARHYNCHUS SCHLIEBENII MANSFELD

Small epiphytic orchids are often very difficult to distinguish when they are growing amongst greyish green lichens on the branches of forest trees. Sometimes, if the collector is in a hurry or has his mind partly on other matters, they can even be collected quite unknowingly. When sorting out the material at home after a collecting trip it is then a considerable surprise to discover that one has found more than one thought.

The miniature plant shown on the next page was collected by accident from the lower slopes of an extinct volcano near Mount Kilimanjaro in northern Tanzania. We were intending to visit a nearby National Park on this occasion and a small herd of elephants had just passed our vehicle on the narrow track when we came to a rough bridge over a stream. Hoping that there were not too many more animals in the vicinity we stopped to search for orchids. The forest which once covered the gentle slopes of the area had been felled a few years earlier and an ugly secondary growth of various bushes had grown up to about elephant height. However, along the stream there were a number of large fig and croton trees which looked a promising habitat for orchids and would provide a useful refuge if the elephants returned unexpectedly.

Within a very short time we had found a healthy clump of several plants of *Aërangis coriacea* which is not very common. These were in the forked base of a small tree growing in the deep shade provided by the fig, and some of the roots must have been dangling in the stream in times of flood.

For some time it seemed that this would be the extent of our collection for the day. But after a while we suddenly began to discover plants of the leafless orchid, *Microcoelia*, previously overlooked on branches which we had already passed by and thought were bare. It was an easy matter to collect some of these on their host branches.

A few days later while tidying up the branches we found another orchid plant growing among the *Microcoelia* and easily distinguishable from it on account of its small leaves. These were a delicate glaucous green in colour, arising close together in the shape of a small fan. The whole plant was less than 5 cm across and the aerial roots had a curiously flattened appearance with very bright green tips. It was carefully detached and tied on to a sturdy piece of rough bark with the hope that it would attach itself to this and produce identifiable flowers.

In less than a year the plant had grown on to the bark quite securely and had produced three sprays of delightful flowers, surprisingly large for its size. Each sepal and petal is approximately 9 × 4 mm and has a scintillating, almost crystalline texture, reminiscent of the surface of frosted glass. The lip is about 6 mm long and has a rich green flash on part of its surface. The club-shaped spur is slightly shorter.

When one examines the small column in the centre of the flower with a powerful lens or dissecting microscope one appreciates the reason for the almost unpronounceable name given to this genus. Viewed from the side the rostellum projects slightly forward and its tip has two small projections so that it resembles a hammer in shape (from the Greek *sphyra*, hammer and *rhynchos*, snout).

This plant has now grown and flowered well for several years in cultivation, in semi-shade, suspended below the bench in the orchid house. Humidity there rarely falls below 50% and the plant is sprayed with water fairly frequently. Temperatures are probably very similar to those in its native habitat, about 50°F at night and rising to 80 or 90°F every day in the hot season and to 70°F or more during the cooler part of the year.

SPHYRARHYNCHUS SCHLIEBENII PLATE 42

TRIDACTYLE BICAUDATA (LINDLEY) SCHLECHTER

The greatest development of orchids with a monopodial type of growth is found in the tropical countries of the old world, from Africa eastwards to some of the Pacific islands. Such orchids are nearly always epiphytic in habit, although the plants may start off at ground level, and the main stem grows steadily on, year by year, bearing the flowers on lateral branches. A number of side shoots which grow continuously from their apex are also often produced. *Vanda* and allied genera in Asia and *Angraecum*, *Aërangis* and their relatives in Africa are perhaps the best known genera with this simple mode of growth.

Tridactyle is another African genus of this kind, and one of the commonest species which is extremely easy to cultivate is *T. bicaudata*. In the wild state it is found in a great variety of habitats ranging from the temperate forests of the southern Cape Province as far north as the coastal forests of Kenya and up to 7,000 feet inland, and westwards across the continent to Ghana. It is often found in large straggling clumps, the lower parts of the stems bare and woody and flowers and green leaves restricted to the terminal few inches, on exposed rocks or as an epiphyte on the lower branches of forest trees. In dry situations it is usually found in deeper shade than in more humid places.

From this brief survey of its ecology it is obvious that this is a very adaptable species and it can accommodate itself to the conditions provided by most orchid collectors. The larger plants are usually pendent so they are best grown on a raft or in a basket which can easily be suspended but small plants which remain upright do equally well in pots of a suitably fibrous compost or bark.

About thirty five species of this genus are known from different parts of Africa and all have the distinctly three lobed lip indicated by the generic name. Very often, as in the species illustrated here, the side lobes of the lip are further divided or fimbriated. Some species bear small groups of flowers which are almost sessile on the woody stem while in others the flowers are arranged in racemes, which are usually slightly shorter than the leaves, from the old and young parts of the stem. In a few species the flowers are white but mostly they vary from pale green or buff to various shades of apricot or orange. Usually they become darker as they fade and most of them have a strong coumarin scent which may persist even when the flowers are almost dead.

T. bicaudata usually has eight to sixteen flowers on each inflorescence and they are borne in two alternating rows on each side of the peduncle with their pedicels curved so that they all lie in one plane. Although each flower is only 2 cm long the plants are very floriferous. Well grown single stems may bear as many as thirty racemes at each flowering while specimen plants with several woody stems are a remarkable sight when they bear several hundred or even over a thousand flowers.

TRIDACTYLE BICAUDATA

PLATE 43

VANILLA HUMBLOTII RCHB.F.

The genus *Vanilla* contains about a hundred species distributed throughout the tropical and subtropical regions of the world. All the species have developed liane-like stems which trail or climb for a distance of several metres through other vegetation but there are two quite different types. The commercial species and *V. polylepis*, illustrated on the following page, both have a shiny elliptic leaf at each node along their glossy green stems. Some other species are completely aphyllous, a greenish or brownish triangular scale being all that exists in the way of a leaf.

Vanilla humblotii is one of the latter group. Its liver coloured stems are about as thick as a man's thumb, and are almost bare apart from the aerial root which appears at each node opposite the tiny scale leaf.

This species is endemic to the Comoro Islands, northwest of Madagascar. Along the western shore of the island of Grande Comore it occurs in a situation which one would have thought was quite unsuitable for an orchid plant. At the edge of fairly recent lava flows which are still black and bare of vegetation are thick tangled thickets of *Rhus* and other bushes, usually only about 2 m high. Temperatures in this area are uniformly high and the insolation is intense. The humidity is also high for most of the year and at times there are very strong winds. Nevertheless plants of *Vanilla humblotii*, which climb through the thickets, occur very near to the seashore and several hundred feet higher, but always in rocky, bush covered country.

Examination of the plant reveals that it is well adapted to cope with these unusual conditions. The aerial or adventive roots which protrude from each node have developed two distinct functions determined by their position on the plant in relation to its habitat. On the upright part of the stem these roots are short and are apparently only used to support the plant and its flowering branches in the host bush. Where the scrambling stems are prostrate the roots have become long, branching, and absorptive with a thick covering of short white hairs wherever they find accumulations of leaf mould between the cracks in the lava rock. Down among the branches of the tangled thicket and protected by its canopy the orchid plant has a shady and humid micro-habitat to which it is well adapted. The additional amenity of strong light, which appears to be necessary for flowering in so many *Vanilla* species, is easily available to these plants when they push their flowering branches through the leaves of the bushes.

The canary yellow flowers of this species open one at a time on the axillary, corymbiform inflorescence, which is about 30 cm long and bears upwards of a dozen flowers. Each of these is rather flat, up to 15 cm across and only slightly shorter. The lip is the largest part of each flower and a very striking feature. It completely encircles the yellow column and has an attractively frilled edge. In its throat it has a deep chestnut brown patch which bears short brown papillae and long rosy red hairs. These hairs are particularly numerous below and around the column and as many of them are over 1 cm long it seems possible that they may have a function in trapping and holding insects for pollination

VANILLA HUMBLOTII PLATE 44

VANILLA POLYLEPIS SUMMERHAYES

Compared with the other large families of flowering plants the orchid family has very few genera of economic importance. The chief exception is *Vanilla*, which is widely known in the form of a spice or flavouring, although the information that culinary vanilla is the fruit of an orchid comes as a surprise to most people.

The Mexican species, *Vanilla planifolia*, is the chief commercial source of vanilla pods and of vanilla essence. Plantations of the crop have been established in other tropical countries where there are frequent but not excessive rains, a high temperature and humidity and an absence of cold winds. Madagascar and the Comoro Islands still produce a very high proportion of the world's supply and the plantations on these islands are humid places with a luxuriance of thick green growth. On the sloping rocky ground in which they are grown *Jatropha* bushes and small trees provide shade and support for the vine-like growth of the vanilla plants. Small aerial roots exserted opposite to the shiny elliptic leaves help the plant to cling to its host but the richly branching absorptive roots are found at ground level in a thick layer of decaying humus.

Vanilla polylepis is a similar leafy plant which has been found in Kenya, Uganda, Congo, Zambia and Rhodesia. It is a plant of humid forests and is usually confined to the banks of rivers, often in rocky gorges. The swollen stem is a rich dark green and bears a succulent, elliptic or ovate leaf at each node opposite to a thin adventive root. The leaves are very thick and have a shiny surface. The channelled midrib is protracted into a small acumen so that heavy rain or the spray from the nearby river will run quickly off the leaf. The plant often grows to a length of several metres before producing flowers on small side branches. For healthy growth the body of the plant appears to need very high humidity and deep shade; but for the production of flowers more light is required. It appears that these are often delayed until there is a gap of some kind in the canopy of the forest and more light can penetrate. When a flowering branch is produced the flowers open successively. Each one lasts only a few days, but the same inflorescence will go on producing buds in successive seasons if its growing tip is undamaged.

Many of the *Vanilla* species have yellowish or greenish flowers and all are distinguished by the lip which completely encircles the column. The flowers of this species are about 5 cm in diameter and their delicate coloration is enhanced by the decorative lip with its rosy mauve flush.

The chief difficulty in growing this and other *Vanilla* plants in cultivation is one of space. The base of a *Vanilla* cutting needs to be provided with a rich, moist, but well-drained leaf mould in a box or tub allowing plenty of room for the absorptive roots. The leafy vine, or the rather snake-like growth of the leafless species, can then be trained over a trellis, or against a wall, in a well-shaded position. In humid tropical gardens they grow very well out of doors and several plants can make an interesting cover for the bare trunk of a tree in a shady place.

VANILLA POLYLEPIS PLATE 45

Bibliography:

Copley, G. C., Tweedie, E. M. and Carroll, E. W. (1964). *A Key and Check List to Kenya Orchids.* J.E. Afr. Nat. Hist. Soc. **24**, (108) 1–58, and **24**, (109) 85–91.

Morton, J. K. (1961). *West African Lilies and Orchids.* Longmans, London.

Perrier de la Bathie, H. (1939). In *Flore de Madagascar*, ed. H. Humbert. **49***e Famille, Orchidées.* Tome **I**.
....................(1941). *Orchidées.* Tome **II**. Tananarive, Imprimerie Officielle.

Piers, F. (1968). *Orchids of East Africa.* Second Edition. J. Cramer, Lehre.

Schelpe, E. A. C. L. E. (1966). *An Introduction to the South African Orchids.* Macdonald, London.

Summerhayes, V. S. (1968). In *Flora of West Tropical Africa*, ed. J. Hutchinson and J. M. Dalziel. Second edition revised and edited by F. N. Hepper. *Orchidaceae* in Volume III, part I. Crown Agents, London.
....................(1968). In *Flora of Tropical East Africa*, ed. E. Milne-Redhead and R. M. Polhill. *Orchidaceae (Part 1).* Crown Agents, London.